Geology of the Florida Keys

UNIVERSITY PRESS OF FLORIDA

Florida A&M University, Tallahassee
Florida Atlantic University, Boca Raton
Florida Gulf Coast University, Ft. Myers
Florida International University, Miami
Florida State University, Tallahassee
New College of Florida, Sarasota
University of Central Florida, Orlando
University of Florida, Gainesville
University of North Florida, Jacksonville
University of South Florida, Tampa
University of West Florida, Pensacola

Geology of the Florida Keys

Eugene A. Shinn and Barbara H. Lidz

University Press of Florida

Gainesville · Tallahassee · Tampa · Boca Raton

Pensacola · Orlando · Miami · Jacksonville · Ft. Myers · Sarasota

Frontis: Jeff Schmaltz, MODIS Rapid Response Team, NASA/GSFC

Printed in Korea on acid-free paper

This book may be available in an electronic edition.

23 22 21 20 19 18   6 5 4 3 2 1

Library of Congress Cataloging-in-Publication Data
Names: Shinn, Eugene A., author. | Lidz, Barbara H., author.
Title: Geology of the Florida Keys / Eugene A. Shinn and Barbara H. Lidz.
Description: Gainesville : University Press of Florida, 2018. | Includes
   bibliographical references and index.
Identifiers: LCCN 2017030447 | ISBN 9780813056517 (cloth : alk. paper)
Subjects: LCSH: Geology—Florida—Florida Keys. | Coral reefs and
   islands—Florida—Florida Keys.
Classification: LCC QE100.F56 .S55 2018 | DDC 557.59/41—dc23
LC record available at https://lccn.loc.gov/2017030447

The University Press of Florida is the scholarly publishing agency for the State University System
of Florida, comprising Florida A&M University, Florida Atlantic University, Florida Gulf Coast
University, Florida International University, Florida State University, New College of Florida,
University of Central Florida, University of Florida, University of North Florida, University of South
Florida, and University of West Florida.

University Press of Florida
15 Northwest 15th Street
Gainesville, FL 32611-2079
http://upress.ufl.edu

CONTENTS

FIGURES

TABLES

ACKNOWLEDGMENTS

The research and results described in this book were influenced and aided by far too many friends, students, and colleagues to list completely here. Although the book is dedicated to Robert N. Ginsburg, one of the modern grandfathers of Florida-Caribbean coral reefs and carbonate geology because of his earlier work and guidance, colleagues such as Paul Enos, Randolph Steinen, Ronald Perkins, Peter Rose, Duff Kerr, and Jim Rodgers were also close influential associates and teachers while Shinn was an employee of Shell Research. Upon leaving Shell in 1974, Shinn was approached by the USGS, especially by then-Branch Chief Peter Rose, to establish a field station on the University of Miami's facility at Fisher Island. The USGS Fisher Island Station began with a small staff of four: Gene Shinn, Harold Hudson, Robert Halley, and Barbara Lidz. Daniel Robbin joined the team later, as did Jack Kindinger. From time to time, the station hosted many researchers, including National Association of Geology Teachers (NAGT) fellowship recipients Lee Kump, Carol Lee, Van Mount, and Bill Zempolich. The NOAA Marine Sanctuary Program, Mineral Management Service, and the National Park Service often provided funding for specific research projects. Throughout this early period, Captain Roy Gaensslen and his shrimp-trawler research vessel *Sea Angel* were key in making it all work. In later years, we would turn his air-conditioned charter yacht *Captain's Lady* into a research and fieldtrip vessel.

During summers, scuba-diving students provided by Our World Underwater, an organization based in Chicago, frequently aided the team. These smart diving students included Valerie Paul, Steve Early, Donna Schroeder, and Adam Ravetch, all of whom went on to distinguish themselves in various technical careers. At times during the 15 years at Fisher Island Station, we hired temporary helpers such as Walter Charm, Linda Sheetz, Phillip Shea, Frank Spicer, Charles Sherrer, and Jeep Johnson. In addition, Lisa Robbins joined us while on a National Research Council postdoctoral fellowship. Professor Randolph Steinen (and students) from the University of Connecticut-Storrs was a frequent field and laboratory volunteer. Duncan Sibley joined the team during sabbatical leave, as did several foreign geologists. Noel James from Canada and Wolfgang Schlager and Peter Swart of the Comparative

Sedimentology Laboratory provided guidance, and the team was especially fortunate to have Bonnie McGregor on board for a year and to teach mapmaking. Bonnie went on to become associate director of the USGS. The Fisher Island Field Station, initially part of the Oil and Gas Branch, was later shifted to the Marine Branch and was supported by then-branch chiefs Richard Mast and Peter Scholle. Throughout it all, we were fortunate to benefit from longtime associations with scuba-diving colleagues and geologists Bill Precht, Robert Dill, and submersible pilot Richard Slater, who facilitated success in our deepwater work through use of the research submersible *Delta*. The complete list of early associates is far too long to incorporate here.

In 1989, Hudson joined the Florida Keys National Marine Sanctuary office in Key Largo, while the remainder of the team including Spicer transferred to the new USGS facility in St. Petersburg. Bonnie McGregor and Peter Betzer were instrumental in facilitating that transition. From that base, we continued research in the Keys, where issues of sewage disposal continued to increase with population growth. New additions to the group were Christopher Reich, Donald Hickey, Ann Tihansky, Virginia Garrison, Chuck Holmes, and at times dozens of others. African dust research was enabled with the microbiological work of Dale Griffin and Christina Kellogg. Fieldwork in the Keys was greatly expedited with Gaensslen's new vessel *Captain's Lady*. We were often aided in the field by professors and students from the University of South Florida's College of Marine Science. Albert Hine along with Stanley Locker and their students were especially important to our success. Peter Betzer, then-dean of the College of Marine Science, was influential in making our relationship with the University of South Florida St. Petersburg successful.

The manuscript for this book was greatly improved by the thoughtful and constructive reviews of Drs. Eugene Rankey of the University of Kansas and Donald McNeill of the Rosenstiel School of Marine and Atmospheric Science, University of Miami. Gratitude is extended to Jim Pitts of the Department of Transportation in Tallahassee, Florida, who kindly supplied the 1975 aerial photographs displayed in this book. Although only a select few of the black-and-white photos are reproduced here, all provided were instrumental in constructing the benthic habitat map (fig. 2.4). In two cases (i.e., figs. 3.3A and 3.13B, which respectively present the extent, Keys-wide, of a nearshore rock ledge and the actual number, four, of outlier-reef tracts), the photos yielded the only concrete evidence for parts of the geologic record that would otherwise have been unknown from other data sets. Appreciation is extended to Russell Peterson, who converted acreage into area for habitat data shown on the pie chart (fig. 2.4), to digital cartographer Lance Thornton, who digitized the final hand-drawn contours for the bedrock and isopach maps (figs. 2.1, 2.2, and 4.2A-B) into the ARC/INFO GIS program, and to graphics artist Betsy Boynton, who rendered all figures and illustrations in format suitable for publication. The authors also benefited greatly from discussions with numerous colleagues through the years but in particular with Ilsa Kuffner and Jim Flocks toward completion of the book. The 1997 seismic geophysical data set was acquired under Florida Keys National Marine Sanctuary Permit FKNMS-27–97. All products using or referencing any portion of the 1997 data set are attributable to and should acknowledge that permit number. We thank Billy Causey, then-Marine Sanctuary manager, for arranging the permit.

ABBREVIATIONS

AAPG	American Association of Petroleum Geologists
ATRIS	Along-tract reef-imaging system
^{14}C	Carbon-14
CIMAS	Cooperative Institute for Marine and Atmospheric Science
CO_2	Carbon dioxide
$\delta^{18}O$	Delta value, the difference between the isotope ratio in a sample and that in a standard, divided by the ratio in the standard, and expressed as parts per thousand per mil
DDT	Dichlorodiphenyltrichloroethane, an organochloride
DNA	Deoxyribonucleic acid, a type of molecule
FIO	Florida Institute of Oceanography
FKNMS	Florida Keys National Marine Sanctuary
GPS	Global Positioning System
GSA	Geological Society of America
Ma	Mega-annum; conventional term for geologic time, an age or point in time in millions of years (= years ago, years old)
MIS	Marine Isotope Stage
NAGT	National Association of Geology Teachers
NOAA	National Oceanographic and Atmospheric Association
PDB	Pee Dee Belemnite, a standard for stable carbon-13 and oxygen-18 isotopes
SEPM	Society for Sedimentary Geology
USFSP	University of South Florida St. Petersburg
USGS	U.S. Geological Survey
UTM	Universal Transverse Mercator coordinate system UV ultraviolet
ybp	years before present

sp.	1 species
spp.	>1 species
cm	centimeters
ft	feet
in.	inches
ka	kilo-annum; conventional term for geologic time, an age or point in time in thousands of years (= years ago, years old)
km	kilometers
m	meters
mi	miles
mμ	micrometers
mm	millimeters

Geology of the Florida Keys

Introduction

The purpose of this book is to examine the ongoing geologic processes inherent in the evolutionary history of the Florida Keys over the past approximately 125,000 years. In language as nontechnical as possible, supplemented by notes where necessary, we discuss the processes that created the coral reefs, lime mud, and ultimately the limestone found in the geologic record anywhere on our planet. Included are significant unanswered questions regarding the geology and biology of the Florida Keys that still need research to resolve.

The book is a virtual guide for those interested in natural history, especially for the carbonate biologists, geologists, sedimentologists, and students who frequent the Keys in great numbers. The Keys and their coral reefs have long provided a premier natural laboratory for scientific investigations of modern and ancient limestone and sedimentary environments. Our focus is primarily on those areas made famous beginning in the 1950s by Robert N. Ginsburg, Twenhofel medalist[1] and considered the father of modern comparative carbonate sedimentology. They remain classic areas for field courses centered on sedimentary processes that aid in understanding formation of ancient limestones anywhere on Earth.

Geologic Context

In spite of, or because of, the uniqueness of the Keys, their reefs and geology have long attracted scientists of many disciplines. As early as 1905 to 1939, a world-famous research laboratory operated on a remote island in the Dry Tortugas, an atoll-like group of islands west of Key West. In addition, because of natural diversity, the Keys hosted many scientific studies during the 1900s. Their distinct geology (composition) and geometry (orientation) allow the island chain to be divided into the lower, middle, and upper Keys (fig. i.1). Based on historic events, often shipwrecks, many Keys coral reefs have acquired individual names. Some are also named after people. In recent years, the authors had the honor of formally designating such a reef in the

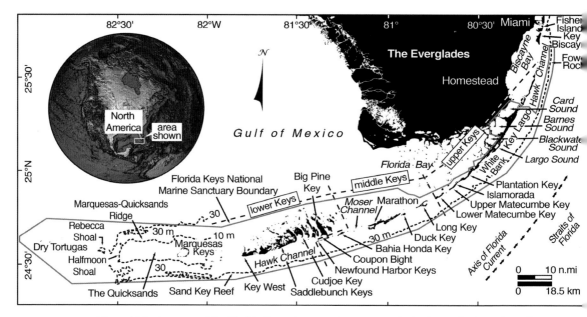

Figure i.1. Index map of the Florida Keys shows major geomorphologies and orientation of the reefal upper and middle Keys and parallel alignment of the linear oolitic bars of the lower Keys. Labeled sites are referred to in text, notes, or figure captions.

upper Keys.[2] At the time (2002), the reef patch consisted of large, pristine heads of *Monstastraea annularis*.

In the following pages, we discuss the geomorphology and limestone origin and composition of major land- and seascape features in the Florida Keys. Because details of the origin and geology deep beneath the Keys are complex and beyond the scope of this book, we focus mainly on the more recent part of their geologic past, specifically, the last approximately 130,000 years of time. This interval comprises the later

Table i.1. Nomenclature for the most recent intervals of the geologic time scale as reported by the International Commission on Stratigraphy

Cenozoic	
Quaternary Period	
Holocene Epoch	4th interglacial stage
Pleistocene Epoch	
Wisconsinan	4th glacial stage
Sangamonian	3rd interglacial stage
Illinoian	3rd glacial stage
Yarmouthian	2nd interglacial stage
Kansan	2nd glacial stage
Aftonian	1st interglacial stage

Note: Note the four cycles of glacial and interglacial stages for North America.

part of the Pleistocene Epoch known as the Ice Ages, and the most recent epoch, called the Holocene. The Ice Ages, or global intervals of glacial stages, were separated by intervals of much warmer (interglacial) phases (table i.1). The Pleistocene began approximately 2.6 million years ago and was followed by the Holocene. According to earlier literature, the Holocene is said to have begun 10,000 years ago, but the interval is now considered to be about 11,500 years old. Many geologists regard the Holocene as simply a continuation of the Pleistocene, though it can rightly be called the era of modern man. The Florida Keys islands on which we walk and live were formed during the later part of the Pleistocene about 125,000 years ago, when sea level was higher than at present. The mantra of the geologist is, "The present is a key to the past." This book demonstrates how and why this mantra is so true and so important.

Social History of the Keys

The Florida Keys are unique, having once been home to several Native American tribes, among them one called the Tequesta. When the Spanish conquistadors discovered the Keys during the sixteenth century, they conquered the Tequesta who had existed there for the preceding 2,000 years. Many were taken to Cuba for conversion to Christianity and where most died. The Tequesta people are no more, but their shell mounds and pottery shards can still be found throughout the Keys and south Florida. Co-occurrence of the mounds and artifacts with intact shells of the gastropod *Cittarium pica* provided important clues to the type of Tequesta lifestyle.[3]

Spain ceded Florida to Britain in 1763, repossessed Florida in 1784, and again ceded Florida in 1821, this time to the United States. During the American Revolution and the war of 1812 with Britain, many European Loyalists fled to Canada and the Bahamas, where life on craggy limestone islands still under British rule was less than ideal. Some, mostly shipwrights, later migrated to the Keys and Key West, which the Spaniards had named Cayo Huecos[4] (Island of Bones). These early English settlers, known as Conchs (pronounced konks), became "wreckers,"[5] salvaging goods, people, and ships grounded on coral reefs.[6] Those who moved to the island were known as "Freshwater Conchs." Folks born there were "Saltwater Conchs." The Keys and especially Key West share a unique, fascinating, and sometimes bizarre social history ranging from rum and drug smuggling to human trafficking.[7] The arrival at Key West of the Flagler Railroad in 1912[8] and the U.S. Navy during WWII added more character to this most unusual island. To this day, Key West and the Keys remain a magnet for unique individuals, including such people as treasure hunters, smugglers, and retired scientists.

Science Comes to the Keys

Alfred Goldsboro Mayer, who held a degree in engineering, decided to study zoology at Harvard University early in the 1900s. While he was there, Alexander Agassiz, director of the Harvard Museum of Comparative Zoology, invited him to coauthor a book on medusae, the umbrella-shaped cnidarians most commonly represented by jellyfish. In 1904, the Carnegie Institution approved Mayer's proposal to establish a laboratory on Loggerhead Key at Dry Tortugas (fig. i.2) some 70 miles (113 kilometers) west of Key West. Mayer succeeded admirably, attracting noted scientists to pursue important marine research and conduct studies of their own design. The Carnegie Laboratory opened for business on this remote island in 1905. Although Mayer, who had changed the spelling of his name to Mayor, died on Loggerhead Key in 1922, scientists continued to conduct research there until 1939. The studies completed on corals by people such as Alexander Agassiz, T. Wayland Vaughan, John W. Wells, and others are known the world over. Most marine science disciplines practiced today were instituted at the Tortugas Laboratory, and the world's first underwater color photograph was taken there in 1926. In all, 146 researchers worked at the laboratory between 1905 and 1939. Subsequent hurricanes have destroyed the buildings, and only remnants of cement sea tanks remain, along with a memorial monument to Mayor placed there in 1923 (figs. i.3A-B, i.4). Mayor's artist wife designed the

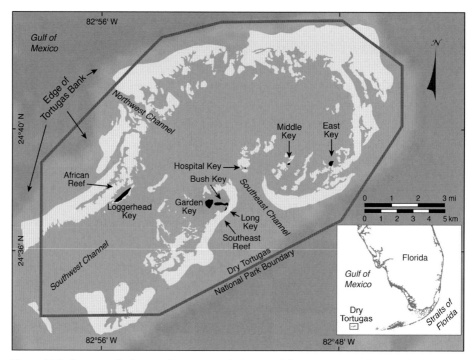

Figure i.2. Index map displays major geomorphologies of the Tortugas Bank. Inset shows location of the Dry Tortugas at the westernmost end of the Florida reef tract. Labeled sites are referred to in text, notes, or figure captions.

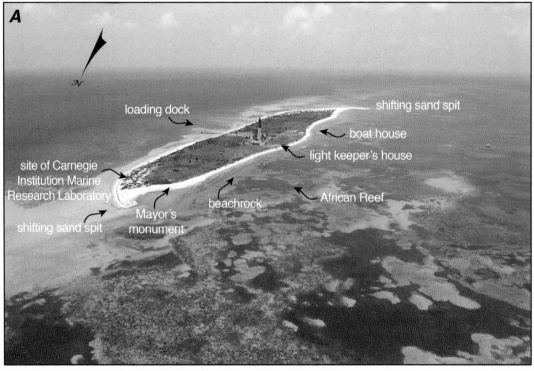

A

loading dock

shifting sand spit

boat house

light keeper's house

site of Carnegie
Institution Marine
Research Laboratory

African Reef

beachrock

Mayor's
monument

shifting sand spit

B

loading dock

shifting sand spit

boat house

light keeper's quarters

beachrock

Mayor's
monument

Site of the Carnegie Institution of Washington
Marine Biological Research Laboratory
at the North End of Loggerhead Key,
Dry Tortugas

Figure i.3. *A*, Oblique aerial view (2004) looking southeast shows Loggerhead Key and the site of African Reef. *B*, Closer oblique view (2004) of Loggerhead Key looking south shows submerged beachrock on west side of island.

Placed here in 1923, a year after his death on the beach near the laboratory, Mayor's monument reads:

Alfred Goldsboro Mayor who studied the biology of many seas and here founded a laboratory for research for the Carnegie Institution directing it for XVIII years with conspicuous success brilliant versatile courageous utterly forgetful of self he was the beloved leader of all those who worked with him and who erect this to his memory

BORN MDCCCLXVIII · DIED MCMXXII

monument, yet she never visited the island. The lighthouse visible in the figures was built in 1858 and at this writing is still operational. Most major lighthouses[9] have been extinguished and replaced by tall, unmanned towers.

Although Darwin in 1842 was arguably the first to study coral reefs in general, it was Alexander Agassiz who conducted studies of the Keys reefs prior to establishment of the Carnegie Laboratory. Because of the toll on ships that wrecked on the reefs, he studied coral reefs at the request of the Light House Service. One of his conclusions was that the reefs were too extensive to be removed! So he recommended construction of steel lighthouses to warn ships of danger. Coral reef science was thus born out of economic concerns to avoid wrecked ships. Although modern technology has reduced the number of shipwrecks, economics still play a role in reef research. Today coral reefs are magnets for tourist and sports-diver dollars.

Economics would drive reef science anew in the 1940s and 1950s, leading to an invasion of geologists and biologists in great numbers. Petroleum geologists stimulated research on coral reefs beginning in the 1950s. It was widely believed that knowing and understanding how modern reefs were initiated and were able to sustain growth would help predict the locations of ancient deeply buried hydrocarbon-bearing reefs. This notion followed the ingrained theory that the present is a key to the past.

Changing Demographics

With the beginning of World War II and an enlarging naval facility, a 12-inch-diameter (1 foot, or 30 centimeters) iron water pipe was constructed alongside Highway US 1 to import freshwater to Key West. The water came from shallow wells in the Biscayne Aquifer south of Homestead (fig. i.1). Before pipeline construction, the city and the rest of the Keys depended on rainwater stored in cisterns. In Key West, shallow wells produced barely drinkable brackish water suitable mostly for washing clothes. The freshwater floating on denser saline water throughout the Keys is thin and variable, depending on rainfall. Climate in the Keys had changed. Before the 1900s, solar salt was commercially produced in salt ponds on the south side of Key West, but wetter weather around the turn of the century prevented continuation of salt production.

WWII, the Navy, and Freshwater

With the coming of freshwater and the navy, including a Naval Air Station, Key West began to prosper like never before; however, communities along the way also tapped the water supply, so that at times only a trickle reached post-WWII Key West. Lack of water severely checked further tourism and population growth. One was lucky to have enough for a shower in the few Keys motels that survived the 1950s and 1960s. Lack of water and presence of only one gasoline filling station between Homestead and Key West, in Marathon (fig. i.1), slowed the flow of tourists. These deterrents vanished in the 1970s with installation of a 36-inch-diameter (1 meter) water pipe. Modern wider bridges that one travels today also aided development. With the arrival of abundant water, motels and housing projects flourished. Duck Key, the town of Key Colony, and other developments, constructed mainly from lime mud and sand dredged from Florida Bay, had their beginnings in the 1950s.

Yet another important factor aided building in the Keys—aerial mosquito spraying! Old DC-3 aircraft zoomed at treetop level, spraying pesticides mixed with diesel oil. Clouds of mosquitoes had always been a deterrent to tourism and population growth. Only the hardy well-tanned Conchs could survive the swarms. It was often said their darkened skin was like leather. Pale-skinned Yankees were fresh meat to the stinging hordes. Mosquitoes, lack of water, and the threat of hurricanes had successfully kept the human population in check, but by the late 1970s, everything changed yet again. The 1970s were the principal drug-smuggling years. Many struggling fishermen would soon be piloting "go-fast" speedboats and driving expensive Mercedes-Benz sedans. As a side effect of their newfound wealth, fishermen began decimating the lobster population as they purchased thousands of additional lobster traps. The traps made them look legitimate. Lobster boats and go-fasts were vital for offloading square groupers (local jargon for bales of marijuana) from mother ships offshore. Jimmy Buffett's songs *A Pirate Turns Forty* and *Margaritaville* popularized this period.

Invasion

In 1980, a human invasion floated in from the south—the Mariel Boatlift. Cuba's economy had dipped, and tobacco crops had succumbed to disease. The United States was falsely accused of practicing bioterrorism on the crops. The result? In April of that year, Cuban President Fidel Castro allowed citizens to emigrate from Cuba at the port of Mariel. He also opened his prisons and mental institutions. Thousands left by the boatloads, and Key West bore the brunt of what became known as the Marielito invasion. The exodus lasted for six months. Demographics of the Keys and south Florida were changed forever.

Oil: In the Keys?

Few Keys residents, especially Freshwater Conchs, know that the Keys were once a focal point for oil exploration. The climax of exploration came in 1943 when Humble Oil Company (now ExxonMobil Corp.) struck oil at 11,500 feet (3,505 meters) near the Everglades town of Sunniland, a small agricultural village near the west coast of Florida (fig. i.5). The oil reservoir is in a Cretaceous (table i.2) reeflike formation that was quickly named the Sunniland Zone. World War II was on, and the country needed all the oil it could find. That black-gold discovery on the edge of the Everglades east of Naples set off frantic searches throughout south Florida and the Keys. Sinclair Oil Company was drilling near Homestead in what is now Cheeca Hammock in Everglades National Park (figs. i.5, i.6). Next was a series of test wells near the intersection of Card Sound Road and the old highway on North Key Largo. Coastal Caribbean Oils & Minerals, Ltd., a Bermuda-based company founded in 1953 by William F. Buckley Sr., had acquired most of the oil rights in Florida waters. Being a small company, they subleased to major companies such as Gulf Oil.

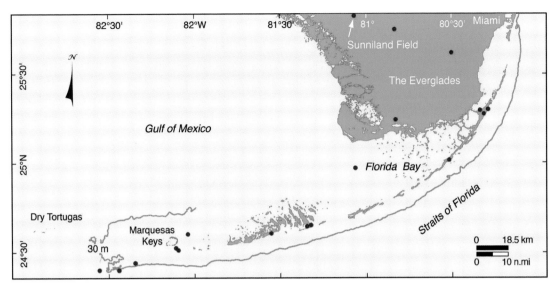

Figure i.5. Index map shows locations of 18 exploratory oil wells (red dots) drilled in south Florida and the Florida Keys between the early 1940s and 1962. Oil was discovered at Sunniland Field in 1943. The Sunniland well is still producing.

By the 1950s and war's end, test wells were drilled on a few Keys followed by a 15,000-foot (4,572-meter) well near Coupon Bight in the Newfound Harbor Keys near Big Pine Key (fig. i.1). Next came a test well about a mile west of Sandy Key in Florida Bay, followed by two test wells drilled in 1959 near the Marquesas Keys 20 miles (32 kilometers) west of Key West. The last test was drilled on the reef tract southwest of the Marquesas Keys in 1960 while Hurricane Donna was ravaging the middle and upper Keys. Two more test wells had been drilled along the outer-shelf reef tract farther west. Oil shows were found in all Sunniland Zone tests, but the amount found was insufficient to justify production at the then-prevailing price—around $5.00 a barrel!

Table i.2. Nomenclature and ages for select intervals of geologic time from the present back to 416 Ma

	Age (Ma)		Age (Ma)		Age (Ma)
Cenozoic Period	65.5–today	**Mesozoic* Period**	251.0–65.5	Paleozoic Period	542–251
Quaternary	2.6–0.01	**Cretaceous**	145.5–65.5	**Permian**	299–251
Holocene	0.01–today	Jurassic	201.6–145.5	Carboniferous	359–299
Pleistocene	2.6–0.01	Triassic	251.0–201.6	Pennsylvanian Epoch	318–299
Tertiary	65.5–2.6			Mississippian Epoch	359–318
Neogene	23.0–2.6			Devonian	416–359
Paleogene*	65.5–23.0				
Eocene Epoch*	55.8–33.9				

Note: * Other names for other periods of geologic time exist between the noncontiguous periods or epochs given. The intervals in bold are mentioned in the text, captions, or notes. Data taken from 2009 Geologic Time Scale.

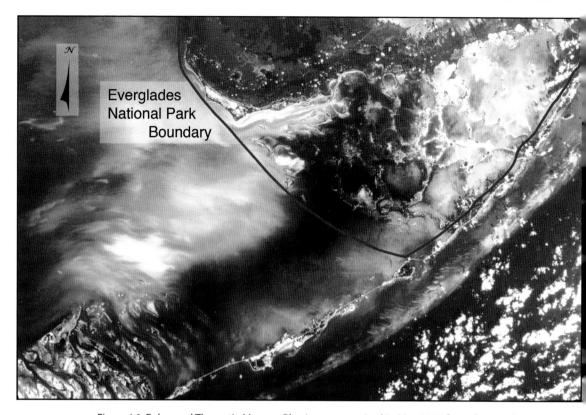

Figure i.6. Enhanced Thematic Mapper Plus image, acquired in May 2000 from the Landsat-7 satellite, shows the four geographic components of the south Florida ecosystem: the Everglades (south part), Florida Bay, the Florida Keys, and the reef tract. Note where approximate southern boundary of Everglades National Park falls. Endangered mangroves fringe virtually every visible shoreline. The upper and middle Keys are an emergent 125,000-year-old coral reef. The lower Keys are emergent fossil-ooid tidal bars of the same age. Seaward of the Keys lie habitats of the inner shelf, Hawk Channel, Holocene sands and reefs on the shallow outer shelf, and 100-foot-tall (30-meter-high) 80,000-year-old fossil reefs on an upper-slope terrace seaward of the shelf edge. The Gulf of Mexico is at left, Straits of Florida under clouds at right. Dense growth of species of the seagrass genus *Syringodium* causes the dark wide band in center.

With establishment of the Florida Keys National Marine Sanctuary and a moratorium on offshore drilling, oil exploration came to a sudden stop. The sanctuary bill was signed on November 16, 1990.

Tourism and Scuba

With completion of the new highway bridges, widened roads, freshwater, and mosquito control, a long-cherished dream for the Florida Keys took root. Tourists came in droves, and with the popularity of a self-contained underwater breathing apparatus (scuba), promoted by Captain Jacques-Yves Cousteau and his television shows, divers arrived and dive shops and charter boats blossomed. In the early 1970s, the first fast-food restaurants took hold, as did liquor stores, adult bookstores, bars, and gee gaw and wooden-toy shops. In the 1950s, other than Overseas Liquors in Marathon, no liquor sources existed between Homestead and Key West. Key West, of course, had always had rum smuggled in from Cuba and elsewhere, as was popularized in Ernest Hemingway's book, *To Have and Have Not*.

The Keys would never be the same. Coral reefs were being loved to death by hordes of snorkelers, scuba divers, and fishermen. Onshore development brought septic tanks, cesspools, and shallow-disposal wells. Contents all went into the porous limestone beneath the Keys.

Everglades National Park was created in 1947 (fig. i.6) at a time before rising concern over coral health. In the 1970s, John Pennekamp Coral Reef State Park was created to protect fish and stop spearfishing off Key Largo. Coral health did not become a serious issue until the late 1970s and early 1980s, when it became clear to scientists that corals were dying. Documented coral death signaled the beginning of efforts to protect the environment from the onslaught of people—and still they came. Plans to control population growth came and went.

The foregoing thumbnail history is intended to prepare visiting scientists and laypeople for a visit to the Keys; however, the real purpose of this book, as stated in the beginning, is to demonstrate the value of the Keys and immediate surroundings as an environmental scientific laboratory for naturalists, geologists, biologists, students, and sedimentologists and to stimulate ideas for further research.

What made the Florida Keys and how did they form? For that information, we go way back before the last Ice Age. It all began many millions of years earlier when the Caribbean was part of a megacontinent called Pangaea, which broke apart to form the Atlantic Ocean. But all that is well beyond the scope and purpose of this book. For that story, we recommend *Geologic History of Florida* by Hine (2013).

1

About the Keys

Processes I

The interaction of numerous and diverse geologic processes over time has sculpted the asymmetric geomorphologies, limestones, and sedimentary environments of the modern Florida Keys. The primary controls on reef distribution and all developmental phases and features of those depositional carbonates were antecedent topography and a fluctuating sea level. Though our focus is on the relatively thin topmost strata (layers) and the most recent part of geologic history—the last 125,000 years, which include the later part of the Pleistocene and the Holocene epochs (table i.2)—deposition in much earlier geologic times resulted in thicker underpinnings. The entire Pleistocene limestone buildup over a 2.6-million-year period is little more than 200 feet (61 meters) thick; the buildup is thickest under the lower Keys and about half as thick under the upper Keys. Below Pleistocene accumulations lies a much thicker Tertiary sequence (table i.2), which in turn rests on thick Cretaceous limestone, deposited during the time of the dinosaurs and accretion of the organic materials that produced the oil-bearing formation at Sunniland. These older rock units deep below Florida and the Keys are also beyond the scope of this book. Again see Hine (2013) for details about these rocks. The younger Pleistocene and Holocene coral reefs, lime sands, and smelly muds of the Florida Keys serve as a natural environmental (biological and geological) laboratory and attract the attention of biologists, geologists, and their students from around the world.

Following discussions of the various Holocene sedimentary processes, spiced with anecdotal stories, facts, and references to previous research, we provide comprehensive maps of the area. The maps are the fruits of extensive research conducted by early workers and most recently by the U.S. Geological Survey (USGS) and others. Over the years, starting in the 1950s, geologists began their pilgrimages to the Keys, often as part of organized water-based field trips. Those classic field trips involved slithering and slogging through smelly mud, slapping mosquitoes, and finally swimming and snorkeling over incredibly beautiful coral reefs. Because of our long history as trip leaders, we provide details and tips on where to go and what to see, in essence a "virtual field trip." We concentrate mainly on "processes" to help put the resultant geology in better context. This virtual trip, based on many years of such field excursions, incorporates contrasting sedimentary processes and environments over a typical 3- to 5-day period. Our virtual trip is suitable for students and professionals, as well as for interested laypeople and naturalists. The reader may then decide to take an actual field trip based on what is presented here. Toward the end of the book, we outline a typical 3-day field trip for those who want the hands-on experience. Be advised, however, that this field of scientific knowledge is itself an ongoing evolving process. Many interpretations of sedimentary processes presented here may be revised by the results of future research. Before we begin, let's start with some basic information about the limestone that forms the Florida Keys and underlies the offshore sediments and reefs.

Corals, Sand Bars, Limestone, and an Impatient Sea

The middle and upper Keys as viewed on maps, satellite images, or from an automobile negotiating traffic on Highway US 1 have a distinctive slightly arcuate northeast-to-southwest profile. These Keys were a living and growing coral reef eventually entombed in reef sand when sea level was 20+ feet (6+ meters) above present (fig. 1.1). Orientation of the lower Keys, however, is clearly different (see figs. i.1, i.6). Beginning at Big Pine Key and extending on to Key West, the highway curves ever more westward, terminating in Key West. Unlike the more linear Keys to the north, the lower Keys were not coral reefs. They were sandbars sculpted by north-south–trending currents and composed of spherical sand grains called ooids.

Ooids and Oolite

Like small pearls, ooids are pinhead-size concretions of precipitated aragonite, the mineral form of calcium carbonate (generally called limestone) that forms mainly in seawater. The majority of limestone on Earth, however, is composed of calcite. Given geologic time, aragonite converts to calcite, especially when exposed to freshwater. Ooids (fig. 1.2) form around a nucleus grain rolling and tumbling in clear current-swept waters supersaturated with calcium ions (Ca^{+2}) and carbonate ions (Co_3^{-2}). When combined, these ions form calcium carbonate ($CaCO_3$). The currents that shaped these sandbars were also responsible for creation of the ooids of which the bars are composed. As with most warm tropical seas, supersaturation and agitation induce calcium carbonate to precipitate. Rolling while precipitating on a nucleus accounts for the spherical egg-shape form—hence the name "ooid." Ooids were once thought to be fossilized fish eggs, which they closely resemble. We are in fact dealing

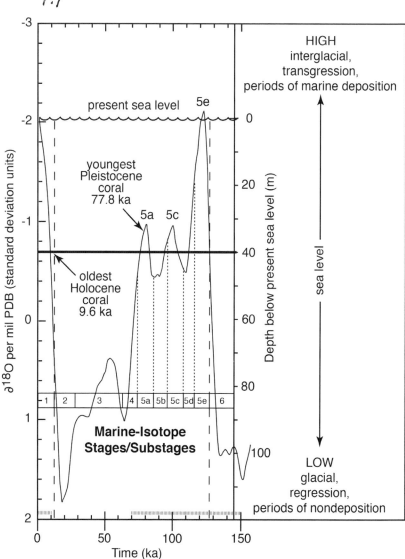

Figure 1.1. Normalized and smoothed marine δ¹⁸O record for the past 150,000 years from Imbrie et al. (1984) with select data from south Florida added. The record reflects changes in sea level in response to waxing and waning of continental ice sheets during glacial/interglacial cycles. Closely spaced tick marks at bottom represent 1,000-year (1-ka) increments. Approximate durations of substages within Stage 5 are, in thousands of years: 5e (127–116 ka), 5d (116–108 ka), 5c (108–96 ka), 5b (96–86 ka), and 5a (86–75 ka). Bold horizontal line represents present depth (131 feet, ~40 meters) of upper-slope terrace as based on actual depths (69 and 102 feet, 21 and 31 meters) of corals flanking a large gap in the depositional record. Those corals have thousand-year radiometric ages of 77.8 ka (Multer et al., 2002) and 9.6 ka (Mallinson et al., 2003) and bracket a period of prolonged exposure. All data, including coral and calcrete ages, point to subaerial exposure of the outer Florida shelf (for 68,000 years, ~68 ka) and the Florida Keys and mainland (for ~115,000 years, ~115 ka) between isotope Substage 5a and the Holocene.

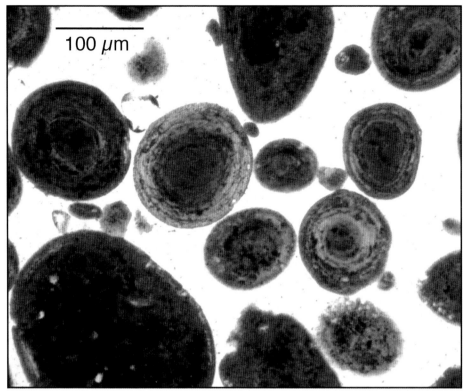

Figure 1.2. Photomicrograph (plain light) shows modern ooids from the Bahamas. The Bahamas and Caicos are the only places in the Atlantic where ooids are forming today.

with what can be called a chicken-or-egg situation. The creation of tidal bars increases current velocity around them, which favors formation of ooids that in turn build the bars, thus further increasing tidal velocity—and so on. Ooid grains may range in size from 125 micrometers (μm) in diameter (smaller than a pinhead) up to 1 millimeter (mm) or more (the diameter of pencil lead). Size is controlled both by water chemistry and agitation. Thus, the grains stop rolling and tumbling with increasing size or are put out of circulation when buried under newly formed ooids.

Water currents tend to sort the grains into uniform sizes and to form distinct layers. The layer sets are composed of alternating accumulations of uniformly sized grains sandwiched between layers with different-size grains. Such layers are seldom horizontal but are inclined at angles determined by current direction. The layers form on the face of ripples and sand waves as the sand bodies migrate across the bottom. While some layer sets may tilt one way, adjacent layers may angle in the opposite direction (fig. 1.3A-D). These distinctly dipping strata are called cross beds, and their presence is a sure sign of reversing tidal currents. Cross-bedded oolitic limestone can be observed in ditches and artificial canals on most of the lower Keys. Cross bedding has been forming somewhere on Earth for many millions (and billions) of years. The next question is, how do these egg-shaped grains become rock?

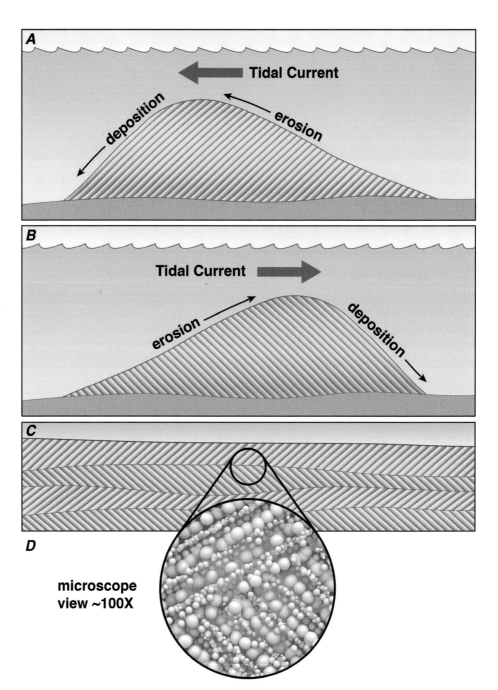

Figure 1.3. Sketches of cross bedding as found in the Miami Limestone. *A-B*, Erosion and deposition alternate on sides of a tidal bar when tidal direction changes, resulting in (*C*) accretion of alternating layers, or strata, of varying thicknesses. *D*, Individual ooid layers also vary in grain size. Used with permission of USGS graphics artist Betsy Boynton.

Once the grains are buried beneath subsequently formed cross beds and cease rubbing against each other, additional aragonite precipitation begins to stick the round grains together—the beginning of becoming limestone. The geologic term for this process is cementation, and once completed, the resulting coarse-grained limestone is called oolite. The lower Keys are oolite, as is the same-age rock beneath Miami and Key West.

Making Oolite

Cementation can take place under seawater but is generally considered more rapid when slightly acidic rainwater percolates into ooid sand and dissolves some of the aragonite. The dissolved aragonite then precipitates out of freshwater as calcite. Calcite is more stable in freshwater than is aragonite; hence, as noted earlier, most limestone on Earth is calcite. During this process, magnesium may enter the calcite crystals to form an even more stable kind of limestone called dolomite. To be more accurate, much of what geologists call limestone in the geologic record is actually dolomite. The precise process by which dolomite forms remains a mystery that has occupied the undertakings of geologists and geochemists for more than a century. Dolomite is limestone that contains magnesium and is expressed chemically as $CaMg(CO_3)_2$ (calcium-magnesium carbonate).

Understanding the process is important because conversion of calcite or aragonite to dolomite increases porosity of the rock and at the same time reduces its solubility. That additional porosity is necessary for the dolomite to contain water, oil, or gas. Thus, a strong economic reason is helpful for understanding how, where, and when dolomite forms.

Although the chemistry is the same, the crystalline form of calcite is distinctly different from that of aragonite. Calcite tends to form blocky or cube-shaped crystals (fig. 1.4A-B), whereas aragonite forms distinctive fibrous or needle-shaped crystals (fig. 1.4C). Even the individual laminations within ooid grains are actually composed of tightly packed needle-shaped crystals lying tangential to the surface of the grains. When ooids cease rolling and remain in seawater, aragonite needles start growing perpendicular to the ooid surfaces. This outward-growing aragonite interlocks and fuses with needles growing outward from adjacent grains. As mentioned, the aragonite-cementing process occurs mainly in seawater, whereas calcite cementation occurs when aragonite sediment is exposed to freshwater. Both processes can occur when water levels change, resulting in both kinds of pore-filling cement between grains. In such cases, the needle-shaped or fibrous cement precedes blocky cement. On the other hand, if ooids are exposed to freshwater, blocky-calcite cement may develop directly on the ooid grains.

Figure 1.4. Photomicrographs (polarized light) show ooids (A) with freshwater blocky calcite, (B) fully encased in calcitic cement, and (C) with acicular aragonitic marine cement. Much of the Pleistocene limestone in the lower Keys and beneath the water and sediment between Key West and the Dry Tortugas is oolite.

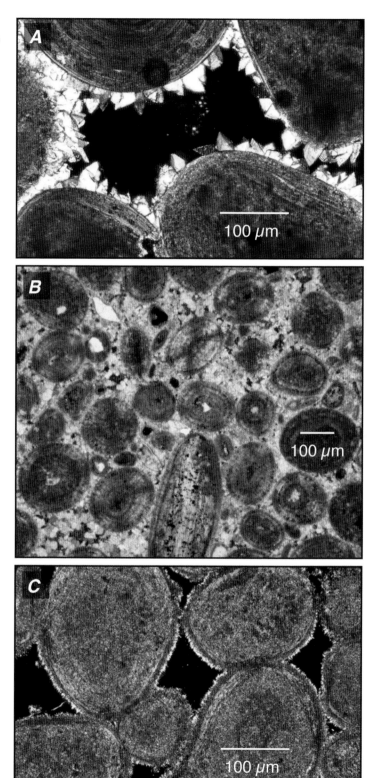

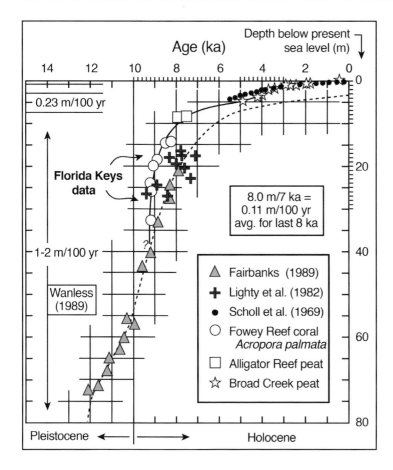

Figure 1.5. Sea-level curve for the Florida reef tract is well constrained by local proxy data (in conventional ¹⁴C ages) as modified (Lidz and Shinn, 1991) from the curve of Robbin (1981, 1984). Data from 8 ka to present are considered reliable for the Florida Keys (solid line). Question mark denotes lack of dates older than ~8 ka because of a lack of *local* material >8 ka (excluding a 9.6-ka date of Mallinson et al., 2003, from Southeast Reef at Dry Tortugas). Green triangles denote data from Barbados (dashed line). Tick marks at top represent 200-yr increments; those at right represent 1-m (3-ft) increments. Rates of sea-level rise shown at left. Upper part of figure shows rise as measured since 1932 by a tide gauge at Key West.

Changes in sea level have been common throughout geologic time. The Pleistocene oolite and other limestones in south Florida were all exposed to freshwater and calcite cementation when sea level was 400 feet (122 meters) below present as recently as 18,000 years ago (fig. 1.5). Diagenesis is the scientific term for any postdepositional physical or chemical change in the rock. Many other forms of diagenesis exist that are beyond the scope of this book.

Before further discussion of coral reefs and oolite, it is imperative to understand that the overriding cause of cementation and diagenesis in the Florida Keys was a fluctuating sea level. Such fluctuations can be caused by tectonic uplift, as in mountain building, subsidence, as in sinking land, or by worldwide changes in sea level. The last is called eustatic change and is generally associated with glacial buildup or melting at the Earth's poles. The graph in figure 1.1 depicts eustatic (global) sea-level fluctuations during the past 150,000 years. The curves are based on studies of microfossil proxy indicators in deep-sea sediment cores and on uplifted coral reef rock and terraces scattered over far-flung areas of our planet.

The major past sea-level fluctuations resulted from both expanding and shrinking of miles-thick glaciers that partially covered both Northern and Southern hemispheres. Glacial ice forms by removal and freezing of water evaporated from the oceans, whereas glacial melting adds water. Water also expands as it becomes warmer. These periods of freezing and melting are known as the Pleistocene Ice Ages, a term conjuring up visions of woolly mammoths, mastodons, and saber-toothed tigers (skeletal remains of which all are found in Florida). To separate glacial and interglacial periods, we use modern terms based on stable carbon- and oxygen-isotope measurements of microfossils in cores of deep-sea sediment. Changes in isotope values occur with changing seawater temperature and salinity. The changes in isotope values are recorded in the calcium-carbonate shells of marine microorganisms that were alive and building their shells at times of warm or cold water—at high or low sea stands. Hence, changing isotope values derived from the microfossil shells serve as proxies for past temperatures and salinities and thus corresponding past relative positions of sea level. A discussion of the origin and use of stable isotopes is beyond the scope of this book. Entire books have been written to explain the origin and use of stable isotopes, and their origin remains an ongoing research discipline in itself.

As stated, we concentrate here on a small fraction of geologic time, the past roughly 125,000 years. This slice of time includes processes similar to those of today, but it was also a time when glaciers had melted and sea level was some 20+ feet (6+ meters) higher than at present. The highest elevation of Pleistocene coral growth in the Florida Keys is on Windley Key (fig. 1.6A-B) at 18 feet (5.5 meters) above present sea level (see *Sea Level Rise in Florida* by Hine et al., 2016). The same-age coral in most of the Bahamas is only about 4 feet (1.2 meters) above present sea level. Does this mean the Bahama Banks platform has subsided, or that the chain of Florida Keys was uplifted? It is a good question to keep in mind. The precise worldwide height of sea level 125,000 years ago varied from place to place, but the overriding changes were global in nature. These differences in sea level in different places remain the source of much study and discussion.

Maximum elevation of the lower Keys oolite islands is around 10 feet (3 meters) above sea level, whereas the same-age oolite underlying the city of Miami is around 20 feet (6 meters) above sea level. Suggestions have been made that the lower Keys have in the past, or are presently, subsiding relative to the upper Keys. Nevertheless, regardless of what actual eustatic sea level was in the Keys during this small slice of geologic time, detailed coring and radiometric dating indicate the upper Keys coral reefs were growing while the lower Keys sand bars were enlarging and being sculpted by tides flowing in and out of the Gulf of Mexico.

Cores drilled into the lower Keys show older coralline limestone underlies the oolite bars. Water-filled channels that separate the lower Keys today were tidal channels when the ooid sandbars were forming. Similar morphology underlies the Miami

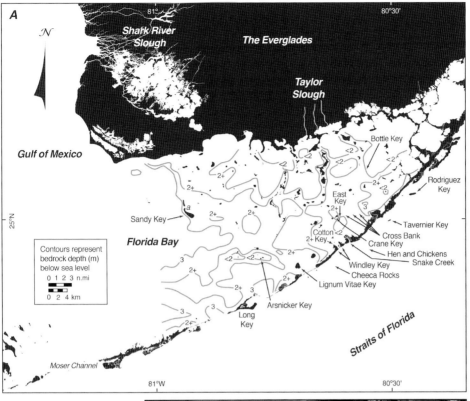

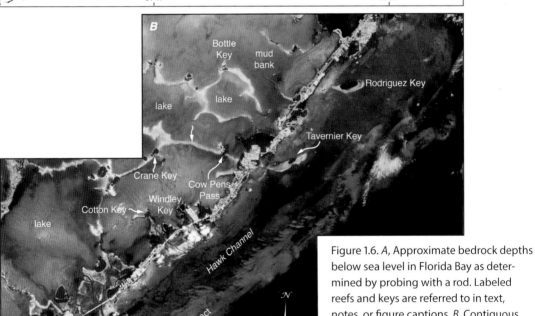

Figure 1.6. *A*, Approximate bedrock depths below sea level in Florida Bay as determined by probing with a rod. Labeled reefs and keys are referred to in text, notes, or figure captions. *B*, Contiguous satellite views of part of upper Florida Keys show major geometries of Florida Bay mud banks and lakes. Note unnamed mud bank (red arrow) recommended for field experience of bay-mud consistency. Also note extents and orientations of banks around Rodriguez and Tavernier keys and the sandy offshore reef tract.

area but is less obvious because the landscape is higher and thus less visible as distinct sandbars. Old tidal channels there have not yet been flooded by seawater. Could this mean the oolite under Miami was uplifted, or has the lower Keys oolite subsided? This question will arise throughout the book.

Figure 1.1 shows that after the high stand of sea level 125,000 years ago (Substage-5e time), the position of sea level fell, not smoothly but in a series of jagged yoyo-like ups and downs that reached its lowest position between 20,000 and 18,000 years ago (Stage-2 time). At its lowest, it was at least 400 feet (122 meters) below present.

Terminology

For the remainder of the book, and beginning with the notes for chapter 1 and the caption for figure 1.5, we switch to conventional scientific terminology and abbreviations[1] and provide measurements in the European metric system followed by English units in parentheses (table 1.1). We also use conventional nomenclature to describe geologic ages and time (tables i.1, i.2; see Abbreviations). Most of the book will be about the Holocene, which began either ~10 ka or 11.5 ka (0.01 Ma or 0.0117 Ma). Relative to the Ice Ages of the Pleistocene and regardless of the actual time of its beginning, the Holocene is the stable period of time when humanity as we know it today reached its zenith.

To describe the various sea-level positions, stands, or stages, we use the formal nomenclature universally employed by oceanographers and geologists, known as Marine Isotope Stages (MISs). These are periods of time determined by differences in oxygen-isotope values preserved in shells, or tests, of deep-sea protozoans called Foraminifera. Note in figure 1.1, which covers only the more recent part of the Pleistocene, highstand stages are represented by odd numbers augmented by lowercase

Table 1.1. Conversion factors for numerical units mentioned in text

To convert	to	multiply by
Fahrenheit	Centigrade	$(0.556 \times °F) - 17.8$
feet	centimeters	30.48
feet	meters	0.3048
inches	centimeters	2.540
inches	meters	2.540×10^{-2}
inches	millimeters	25.40
micrometers μ	millimeters	0.001
miles	kilometers	1.609
millimeters	centimeters	0.1
nautical miles	kilometers	1.853
pounds	kilograms	0.4536
tons	kilograms	907.1847

letters a through e to represent substages. Low stands are represented by even numbers. Thus, the high stand at 125 ka is formally known as Marine Isotope Substage 5e (with a capital S since it is a formal term identified by its number), or MIS 5e. Also note lesser highstand Isotope Substages 5a, 5b, 5c, and 5d. Substage 5e is equivalent to the Sangamonian interglacial stage (table i.1) in older American geologic literature. Two formal limestone formations characterize Substage-5e time in the Florida Keys: the Miami Limestone underlying Miami and comprising the lower Keys, and the Key Largo Limestone comprising the middle and upper Keys. The entire Pleistocene encompasses about 17 isotope stages spanning some 2.6 million years (table i.2).

Sea-Level Ups and Downs

Sea level in figure 1.1 is represented in meters (m) with 0 being its present position. Note that the various ups and downs determined by different authors correspond closely. Some authors have indicated a sea level higher than present during Stage 1. No persuasive evidence for this higher level has been found in the Florida Keys, though there is ample evidence in the Pacific Ocean. Our planet is not a perfect sphere, and its shape changes slightly through time, periodically bulging on one side and compensating by depressing on the other.

Additional details of sea-level positions and fluctuations in Florida since Isotope Stage 2 are provided in figure 1.5. This detailed curve is based on ^{14}C (carbon-14) analysis of local submerged peats and corals that once grew near the sea surface. Specifically, note that the rate of rise slowed significantly between 8 and 6 ka (Wanless, 1989). Also note that Th/U (thorium/uranium) and ^{14}C ages of shallow-water corals from different sites were used to construct the curve. The corals came from a core off Barbados, from a fossil Holocene reef off Fort Lauderdale north of Miami, and from individual corals within a 14-m-long (45-ft) core from Fowey Rocks off Miami (fig. i.1). Though the Holocene officially began at the end of the Pleistocene at around 10 ka, the distinction is a fuzzy one. Many would agree that the Holocene began earlier, at 11.5 ka, or that it is simply a continuation of the Pleistocene.

Holocene reef accretion in the Florida Keys originated around 6 ka when seawater began to encroach on the Pleistocene limestone shelf in areas several kilometers seaward of the Keys. Significant Holocene coral accumulations before 6 ka at depths deeper than 20 m (65 ft) are absent or have not yet been discovered. However, drowned fossil beach dunes composed of oolite, evidence of paleoshorelines, are found off the southwest shelf margin between 61 and 46 m (200 and 150 ft) of water. These sedimentary features date between 14.5 and 13.8 ka (Locker et al., 1996).

Why did ooids form along a shoreline at 13.8 ka when sea level was 46 m (150 ft) below present? No true ooids are forming in the Keys today, yet they are forming abundant shoals near sea level in areas of the Bahama Banks. Why not in Florida?

A possible answer—seawater was more saline and alkaline as a result of evaporation when the glaciers were forming. But why then did ooids form shoals in Florida during the Pleistocene when sea level was very near present level? Will the sea continue to rise and reach its prior level of 125 ka? The latter is an important question for humanity, not just for those living in the Florida Keys but also for coastal populations worldwide! For a highly detailed analysis of the Keys Pleistocene components, two classic works by Multer (1977) and colleagues (Multer et al., 2002) are recommended.

Creation of Florida Bay and Glorious Mud

A few thousand years after corals began growing offshore, a unique environment arose on the other side of the Keys—Florida Bay. Here, we restrict the name of Florida Bay to the shallow triangular lagoon with anastomosing mud banks, lakes, and mud islands between the chain of Keys and the tip of mainland Florida within the boundaries of Everglades National Park (figs. i.6, 1.6A-B). Mud banks exist beyond these boundaries in the area on the Gulf side of the lower Keys as well; however, the triangular area west of the upper Keys has long served as a famous natural laboratory for learning the origin of fine-grained limestone. In fact, what we call mud actually consists of soft sand-size fecal pellets (excrement from marine organisms, mainly worms), but in one's hand it's just gray smelly mud.

Numerous geologists have investigated the sticky, reduced, H_2S-rich (hydrogen sulfide, smells like rotten eggs), silty calcium-carbonate mud since the turn of the century. After early studies (Drew, 1914), the mud was initially called drewite. Ginsburg and Lowenstam (1958) made the area and its drewite/lime mud famous. Cross Bank and Crane Key (fig. 1.6A) will forever be associated with Ginsburg's early studies and the many field trips he led there. One of his disciples, Jerry Lucia, once recorded a song about these trips based on an English show song by Flanders and Swann about the hippopotamus. The title was *Mud, Mud, Glorious Mud*. Ginsburg's PhD dissertation on these sediments led to his employment by Shell Development Co., the research arm of Shell Oil Co., and establishment of a research station in Coral Gables, Florida. Soon petroleum geologists from around the world were descending on Florida Bay to study its mud. Why mud?

Much of the geological record consists of vast quantities of lime mudstone,[2] and abundant oil and gas have been extracted from ancient limestone that was lime mud millions of years ago. Many limestone reservoirs, or aquifers, also supply the freshwater we drink, especially in south Florida.

The mud is smelly for a reason. Research has shown that when heated and compressed under great pressure, the organic matter within the mud is transformed into crude oil or gas, or both. A simple 30-day experiment of heating and compressing Florida Bay mud produced a small amount of tarry crude oil (Shinn et al., 1984). No wonder that geologists from around the world are drawn to the smelly mud! They search for clues to better understand and predict areas on Earth where drilling might hit hydrocarbon pay dirt. Academics have also felt the magnetic pull of the mysteries of the bay. A slog or crawl across smelly Cross Bank in Florida Bay created a lasting memory for thousands of student geologists. Many prospered and literally hit "black gold" thanks to the unique knowledge they gained there.

To understand the origin of Florida Bay, we examine the rise of sea level once again (fig. 1.5). The Pleistocene limestone surface beneath the bay water is relatively flat but is inclined toward the southwest. Near the northeast corner of the bay, near Highway US 1, the underlying limestone lies about 1 m (3 to 4 ft) below sea level. Toward the southwest, the surface lies about 2 to 3 m (7 to 10 ft) below sea level (fig. 1.6A). Its surface has often been likened to a table with one leg shorter than the rest. Except for a small place near Crane Key, the lime rock is flat and composed of pelleted grainstone.

The flat surface of the bay limestone floor has at least one anomalous topographic high located near East Key (fig 1.6A). Coring revealed the high is a Pleistocene coral patch. Because of its elevation, approximately 1.2 m (4 ft) above the surrounding pelleted grainstone, it has not been smothered in lime mud but is instead encrusted with sponges and gorgonians. Another anomalous high exists near Arsnicker Key. Lignum Vitae Key is also an uncharacteristic high composed of Pleistocene coralline limestone. That Key is named after the exceedingly hard wood, *Lignum vitae,* that grows there. Early ships often used that wood for propeller-shaft bearings.

Comparison of bay depths with the sea-level curve in figure 1.5 indicates that the limestone underlying Florida Bay most likely supported a swampy freshwater marsh (similar to the present Everglades) at the same time corals were beginning to grow on the Atlantic side of the Keys. The area remained that way until sea level rose high enough to begin trickling into the bay depression through low-bedrock areas between what we now call the Florida Keys. The flooding of these topographic lows also had significant effects on the developing reefs, as will be discussed later.

Once seawater covered the limestone platform, lime mud began to form and accumulate over what had been marshland. With continued rise in sea level, Florida Bay became ornamented with numerous anastomosing lime-mud banks and low islands that, during winter months, are often slightly awash (fig. 1.7). Some were initiated as mud-shoreline beaches like those presently forming along the northern margin of the bay (Wanless and Tagett, 1989). Typically, most mud banks are attached to a mud-based mangrove island that rests on underlying Pleistocene grainstone. Abundant probing with metal rods and push cores revealed that mangrove-populated mud islands and mud banks generally overlie peat and calcitic mud that cap the Pleistocene limestone. The grainstone is often coated with a reddish-brown laminated layer of limestone called caliche.

Figure 1.7. Oblique aerial view is of storm-flooded islands in Florida Bay. Note milky color of surrounding water, caused by suspended lime mud.

storm waters
flood islands

What is Caliche?

Caliche is generally a western U.S. term for so-called caprock used as "road metal." Caliche occurs the world over and is often called calcrete, or soilstone crust. In the Keys and elsewhere in the Caribbean, caliche consists of a thin, exceedingly dense, laminated, reddish-brown limestone crust a few millimeters to centimeters thick. It forms on limestone elevated above sea level. The crust generally forms beneath a peaty soil and contains voids made by roots and root hairs.

In the Keys, such crusts range from a few millimeters to several centimeters (⅛ inch to several inches) in thickness (fig. 1.8A-E). Because of its density, caliche is impervious and retards downward flow of rainwater. Caliches form when aragonitic mud and sand or dust dissolve in damp organic-rich and acidic peaty soil. As the water moves downward and the soil dries, the dissolved calcium carbonate precipitates as a very thin film of calcite on the underlying limestone. The accretionary layers contain organic material derived from the soil, and the red-to-brown color is due to oxidized iron and clay derived from the atmosphere. The quartz silt and clay along with numerous other elements, some toxic to humans and other biota, arrive in the Caribbean and Florida Keys as dust transported in trade winds from North Africa (Prospero and Nees, 1986; Muhs et al., 1990; Prospero, 1999). Multer and Hoffmeister (1968) described the crusts in detail before they were discovered to contain African dust.

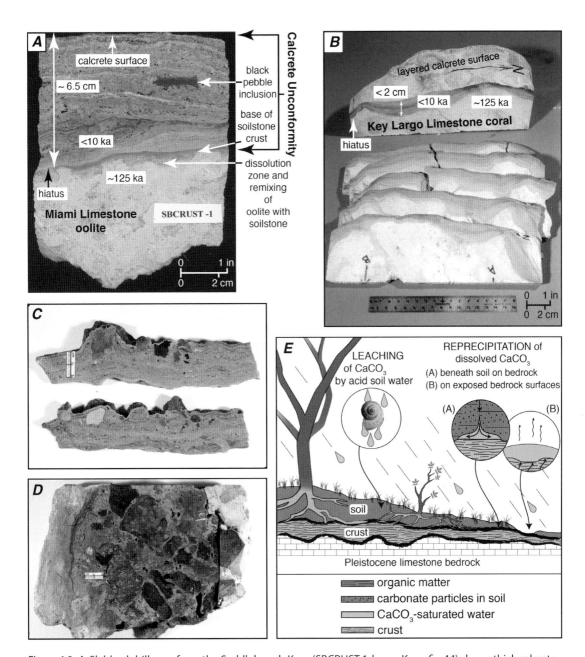

Figure 1.8. *A*, Slabbed drill core from the Saddlebunch Keys (SBCRUST-1, lower Keys, fig. 1.1) shows thick calcrete or caliche overlying Miami Limestone (also called Miami Oolite). *B*, Rock sample from Adams Cut (also called the Key Largo Waterway, upper Keys, fig. 1.1) shows same calcrete capping Key Largo Limestone coral. Crusts that overlie oolite are generally thicker than those topping coral. Oxidized iron from African dust causes the reddish-brown color. The hiatus, or gap, in these rock records represents an interval of more than 115,000 years during which no marine deposition occurred in the Florida Keys. *C*, Large angular naturally blackened pebbles embedded in brown, layered, fine-grained calcrete. *D*, A multicolored, well-cemented, artificially blackened limestone collected ~6 m (20 ft) below sea level from quarry tailings in a solution pit on Big Pine Key (lower Keys, fig. 1.1). Note layered calcrete crust on left edge of specimen. Blackened fragments include preexisting calcrete and fossiliferous Key Largo Limestone. Specimen was heated in the laboratory to reproduce darkening similar to colors of blackened pebbles widely found in Florida Pleistocene and Holocene records. Original rock colors are visible in unheated section at right edge of specimen. *E*, Schematic shows some processes responsible for calcrete-forming subaerial crusts. Graphic modified from Multer and Hoffmeister (1968).

Caliche is a definitive indicator of subaerial exposure (exposure to air) during sea-level fluctuations. As such, its presence indicates periods of marine nondeposition. These periods represent hiatuses or gaps and create unconformities[3] in the rock record. Perkins (1977) used the presence of caliche to correlate various subsurface Pleistocene marine depositional units that comprise south Florida and the Keys. Caliche often contains thin discontinuous black laminations and angular black pebbles that were found indicative of grass and forest fires caused naturally by lightning strikes (Shinn and Lidz, 1988). Shinn et al. (1994) demonstrated that caliche layers are impervious when not fractured and may serve as impermeable seals. These aquatards greatly influence groundwater flow in south Florida and the Keys.

The gray calcitic mud, locally called marl, under many bay islands is similar to that presently forming in the southern Everglades freshwater marshlands. Several species of freshwater snails characterize this marl. Scattered randomly throughout the marshes and marl prairies are vegetated tree islands called hammocks. The trees and bushes that live there contribute to the formation of peaty soil that in turn is needed by the trees and vegetation found there. The surface of the hammock peaty soil is about 30 or 60 cm (1 or 2 ft) higher than adjacent water and marl. Periodic fires in the surrounding sawgrass prairies keep the outward spreading of tree hammocks in check.

Extensive coring of mangrove islands in Florida Bay has shown that the underlying peat and marl are similar to those forming the modern tree hammocks and surrounding marl/sawgrass prairies. Pollen species like those in tree-hammock soils have also been identified in the peat beneath bay islands. Bay mud islands were formed when rising sea level flooded freshwater hammocks, a conclusion supported by ^{14}C dating of the underlying peat and marl (Scholl et al., 1969). Thus, the southern boundary of the Everglades at 4 ka to 3 ka was farther south of its present location before being gradually drowned by rising sea level. Mud-laden beaches and shorelines likely formed the bases of mud banks as the landward edge of the sea moved northward into the Everglades.

Mud Islands and Hurricanes

A long-standing mystery was how mud islands accreted upward and maintained their position above mean water level while sea level was rising. Submarine mud banks have also kept pace with a rising sea. How is that possible?

Hurricanes finally provided the answers. Hurricanes Donna in 1960 and Betsy in 1965 delivered the evidence during storm surges that deposited thick slurries of soupy lime mud on submerged islands. As the storms passed, water levels subsided, stranding several centimeters (inches) of new mud. At the same time, dense growths of red and black mangroves protected island margins while trapping sandy sediment.

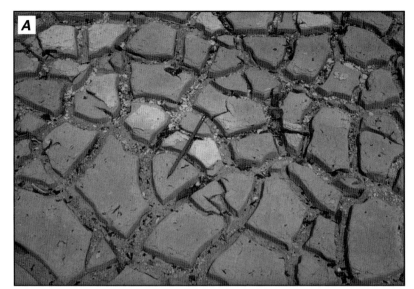

Figure 1.9. *A*, Desiccated mud layer deposited by Hurricane Donna on an island in Florida Bay (pencil for scale). *B*, Laminations within the stiff lime-mud layer.

The new mud settled in island-interior areas that lacked mangroves and supported mainly salt ponds. Once deposited in the ponded areas, the mud layers become hardened and cracked when desiccated by the sun (fig. 1.9A-B). Characteristic mud cracks and air-filled voids called birdseye structures[4] formed within the mud as it dried (Shinn, 1968). Blue-green algae soon created leathery mats over the sun-dried layers. Ginsburg and Lowenstam (1958) have shown how such mats form quickly. Once formed, the surface is ready for the next storm. Push cores (fig. 1.10A-D) reveal this process has been repeated many times over. The addition of new mud layers during hurricanes thus demonstrates how the islands continue to grow upward, keeping them just ahead of the rising sea. The process also produces characteristic air-filled millimeter-size voids indicative of this type of sedimentation. The voids, called birdseye vugs or fenestrae, are recognized in many ancient limestones millions of years in age.

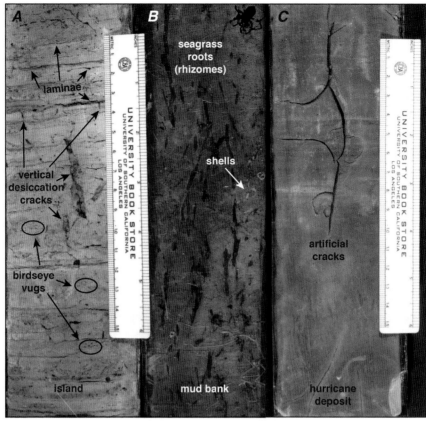

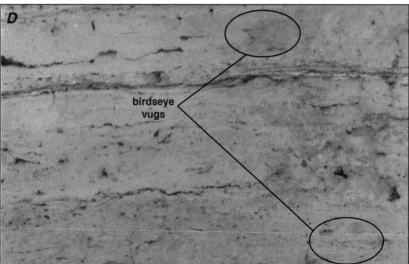

Figure 1.10. Clear polyester-resin–impregnated cores from Florida Bay are from (A) an island, (B) a typical mud bank, and (C) a mudbank hurricane layer. A, Note laminated algal storm layers with desiccation cracks and birdseye vugs, encircled and in close-up view (D). Oxidation causes creamy color. B, Mudbank core contains vertical turtlegrass holdfasts (roots? rhizomes?) and scattered shells. Reduced mud causes gray color. C, Very thin horizontal laminations and a lack of roots characterize a mudbank hurricane deposit. Vertical cracks are artificial, formed during drying prior to impregnating with resin.

Hurricanes also affect mud islands in other ways. While some sides of islands are eroded, other sides accrete and grow laterally. Islands such as Bottle Key (fig. 1.6A) were shown to have migrated southward significantly from their original positions. Ginsburg and Lowenstam (1958) and later Ball and colleagues (1967) studied these processes of upward growth, and Perkins and Enos (1968) documented lateral migration. These studies revealed the importance of storms as agents of sedimentary processes, and the sedimentary structures they produce help identify similar sedimentary features in limestone hundreds of millions of years old. Such identification helps a geologist on a mountainside outcrop determine the origin of the limestone being examined. "The present is a key to the past" at work!

Mudbank Migration

Two distinct features characterize linear mud banks, especially those oriented more or less east-west. Most have a band of coarse-grained sandy/shelly sediment on their north-facing flanks but support only soupy mud and seagrass on their south flanks. Thick meadows of seagrasses populate their flat tops. Much of the sandy material on the north flanks consists of foraminiferal tests, the protozoan genera that create these coarse-sand-size calcium-carbonate grains. Some species of benthic or bottom-dwelling forams, as they are generally called, along with a host of other encrusting organisms, live mainly on the seagrasses that dominate the banks. They and various pelecypods, gastropods, and other shelled molluscans contribute to the sediment upon death and periodic shedding of grass blades.

For years, a mystery was, why are the south sides of the banks composed of mud with the texture of face cream while the opposite sides are shelly? One can walk easily on the shelly side but sink up to one's knees on the middle and south sides, yet the south sides are exposed to prevailing southeasterly winds. Should not the south flanks also be shelly because of exposure to prevailing winds and waves?

Again hurricanes, in addition to winter cold fronts, provided the answer. When Hurricane Donna passed over Key Largo in 1960, the first and strongest winds blew from the north. Winds at speeds up to 241 km/hr (150 mi/hr) churned up waves and blew water out of Florida Bay, lowering the water level by 1 m (3 ft) or more. At that time, waves battered the exposed north-facing margins of mud banks and islands. The edge of the mud bank in essence became a beach for a few hours. During the brief assault by waves, mud was carried away while shells and sandy sediment became concentrated in a thin layer over relatively stiff coherent mud. At the same time, suspended mud was swept across the banks and deposited as slurry on the south leeward flank, where calmer conditions prevailed. In many areas, mud settled and completely buried large areas of seagrasses. As much as 30 cm (1 ft) or more of mud were deposited in a brief few hours. Fortunately for sedimentologists, the mud deposited by Hurricane Donna contains characteristic thin laminations (figs. 1.9B, 1.10C), and the accumulation was too rapid and too thick to be destroyed later by burrowers and by seagrass rhizomes and roots. A core taken in these areas today still contains those and many previously deposited storm layers (fig. 1.11A-B). Cores taken elsewhere, such as from beneath the shelly layer on the opposite side of a mud bank, usually contain the same distinctive laminations. The fact that these distinct nonburrowed

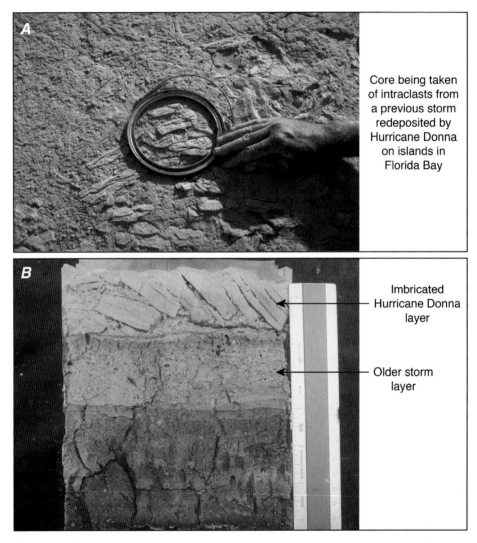

Figure 1.11. *A*, Typical mud cracks on surface of desiccated storm layer. Note imbricated intra-clasts (desiccated storm layers from a previous storm) redeposited by Hurricane Donna on Florida Bay islands. Push core is being taken with a 1-gallon paint can with bottom removed. *B*, Cross section of plastic-impregnated core shown being taken in (*A*). Note imbrication of desiccated mud layers as seen in plan view in (*A*). Older storm layer containing abundant birdseye vugs is visible below the Hurricane Donna layer. Algal layers are interspersed with storm layers farther down core.

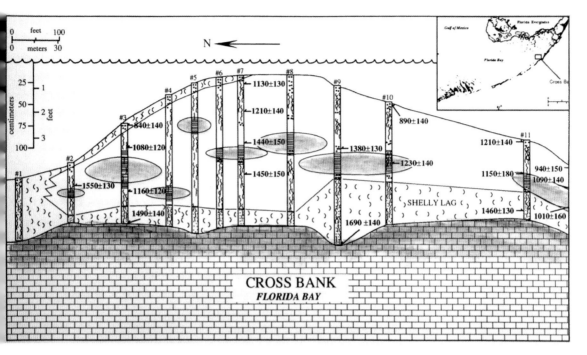

Figure 1.12. Cross section made from 11 cores taken across Cross Bank mud bank in Florida Bay. Shaded ovals indicate laminated storm layers like that shown in figure 1.9A-B. Surface to right of bank is covered with shallow roots of *Thalassia testudinum* (see fig. 1.10B). Surface to left is composed of a shelly lag or grainstone of bivalve pelecypods, or a wackestone to packstone. Migration direction is to the south (right). Erosion is on north side of bank (to left in drawing). Numbers are bulk uncorrected ^{14}C ages. Inset shows site of Cross Bank. (From an unpublished study by Shinn).

laminations formed on the south flanks indicates that some mud banks have actually migrated southward (Wanless and Tedesco, 1993). Transects of cores across a bank further confirmed that mud banks had migrated during previous storms (fig. 1.12).

As the central eye of Donna moved westward, winds shifted and blew from the southwest and west. At that time, water was pushed back into the bay, and the water level rose to about 2 m (7 ft) above normal. During that time, mud was also deposited on the islands, while most of the mud on the south sides of the mud banks remained as the winds waned. These observations after just one hurricane solved a long-standing mystery.

Florida Bay has indeed proved to be a natural laboratory for unraveling past and present secrets of Nature, but what about the future? Clearly, as sea level continues to rise, the past will be repeated. The Everglades marshlands and hammocks will be submerged, and Florida Bay will simply continue to expand. How far it will advance is unknown, but, as it has in the past, the extent will depend at least in part on preexisting topography.

Tidal Passes, Inimical Waters, and Coral Distribution

While Florida Bay and its mud banks were forming, coral reefs were accreting on the other side of the Keys. Before discussing the different reef types, we should note that the reef tract is not a continuous belt of coral. In certain broad areas, reef growth has either been halted or was never initiated. Wide areas of poor reef growth occur mainly in the lower Keys and north of Key Largo off Biscayne Bay. The most notable example that serves as a model is the area seaward of Moser Channel in the middle Keys (fig. i.1). Moser is the wide channel spanned by the well-known Seven Mile Bridge. Other smaller passes occur south of Lower Matecumbe and Long Key as well as toward and beyond the island of Key West. The paucity of reefs seaward of tidal channels has been noted by numerous authors (e.g., Vaughan, 1914; Ginsburg and Shinn, 1964; Enos, 1977) and later elaborated upon by USGS personnel (Shinn et al., 1989; Lidz and Hallock, 2000; Lidz et al., 2007).

In recent years, an explanation for the absence of mature reefs opposite tidal passes became popularized as the inimical-water hypothesis (Ginsburg and Shinn, 1964). This hypothesis has been confirmed by numerous observations of tidal waters flooding over nearby offshore reef areas. Chemistries, nutrient contents, and temperatures of waters from the Gulf of Mexico and Florida Bay are highly variable. These waters may be excessively warm, hypersaline, nutrient-rich, and mud-laden during summer months and more frigid than many coral species can tolerate in the winter. The shallows of Florida Bay are easily heated in summer and chilled during winter months. Farther west beyond the Marquesas Keys, and including the Dry Tortugas, no island barriers exist to protect any coral growth from cold Gulf of Mexico waters during winter months. During extremely cold winters, as observed in 1977 (Roberts et al., 1982), chilled muddy water reached the outer-shelf reef tract. The effects of such events are greatly enhanced when low Atlantic tides coincide with cold winter winds from the north. Water as cold as 9°C (48°F) was recorded by a thermograph in Snake Creek tidal pass (Hudson, 1981). Satellite imaging showed muddy water around 12°C (54°F) flowed seaward over the reef tract in areas opposite tidal passes (Roberts et al., 1982).

During summer months, hurricanes and tropical storms flush offshore reefs with warm muddy water (Ball et al., 1967; Perkins and Enos, 1968). Mud-laden water lay across the reef tract for about three weeks following passages of Hurricanes Donna and Betsy as well as after Hurricane Andrew in 1992. Long periods of muddy water are deleterious to coral growth. These observations indicate that such flooding has occurred many times since rising sea level created Florida Bay. Depth within tidal channels when compared with the sea-level–rise curve in figure 1.5 indicates inimical bay water began flowing via tidal exchange over the fledgling offshore reef tract between 6 and 5 ka.

Figure 1.13. Dense stand of healthy elkhorn coral *Acropora palmata* capped Carysfort Reef prior to the 1980s. White square is ~10 cm (4 in.) across. Courtesy of Dr. Phil Dustan, College of Charleston (Ret.).

Rapidly growing, branching, reef-building corals such as *Acropora palmata* are the least tolerant of both excessively warm and excessively cold waters, whereas massive head or boulder corals, such as species of *Montastraea*,[5] *Diploria*, *Colpophyllia*, and *Siderastrea*, are more tolerant. Temperature and salinity tolerances of most Florida corals were determined long ago by observations and experimentation in the early 1900s at the Tortugas Carnegie Institution Research Laboratory (Vaughan, 1915a,b). Temperature tolerance and growth rate of *A. cervicornis* were also determined in later transplant experiments (Shinn, 1966).

Underlying topography and interplay of both inimical and clear, more stable waters from the Straits of Florida (Gulf Stream) control reef distribution. For example, reefs off Key Largo, the longest island in the Keys and with few narrow tidal passes, are protected from unfavorable bay waters. These reefs are thicker and have more coral diversity than those off the lower Keys, where tidal passes are numerous and wide. This relation, however, became less noticeable with the advent of Caribbean-wide coral diseases and bleaching in the late 1970s and 1980s. Corals at this writing are being impacted throughout the Keys, and differences between older dead reefs opposite tidal passes and those off Key Largo are less obvious now than in the past.

Nevertheless, coring and seismic profiling show that in general Holocene corals along the outer-shelf reef tract facing the Gulf Stream have never reached their full potential. Those techniques indicate that for the most part the reefs are seldom more than 1.8 m (6 ft) thick. Their lack of upward accretion remains a mystery. Some of the reefs along this trend, however, are thicker. The thicker reefs created mainly by rapidly growing *A. palmata* (fig. 1.13) are generally those marked by lighthouses and associated with underlying topographic highs. In those places, coral growth has kept pace with rising sea level, and the reefs are awash at low tide. Some have even created small sandy islands composed of reef sand and rubble. Ships since Spanish and British sailing days and recent modern ships of steel have frequently run aground, and small pleasure boats of fiberglass hit the reefs almost weekly. The loss of valuable cargos and creation of the wrecking industry led to the construction of lighthouses during the mid-1800s.

Surprisingly, the majority of the so-called reef tract, about 98%, lies in 6 to 10 m (20 to 35 ft) of water. Some areas that appear to be coral reefs to the average diver are in fact Pleistocene limestone encrusted with algae, gorgonians, sponges, and the random head of a massive-coral species. They are not true coral reefs, but cores show coral reefs developed there during the Pleistocene. Why then and not during the Holocene?

The named well-marked reefs that have grown to present sea level are on average around 12 m (40 ft) thick. Why the major part of the outer-shelf reef tract adjacent to clear oligotrophic Gulf Stream waters of the Florida Straits has accreted so little is not known. Corals are supposed to prefer warm, clear, nutrient-free, so-called oligotrophic water the world over.

Also surprisingly, the thickest and most mature Holocene reefs are those located 1.6 km (1 mi) or more landward of the outer-shelf reef tract. Topography of underlying limestone, rather than water quality, seems to be the major factor controlling their distribution and growth. Many questions about reef distribution, geometry, and thickness remain unanswered. Why is it that corals that grow upward much faster than sea level has risen during the past 6,000 years have not yet reached the surface? Except for areas opposite tidal passes, why have the corals not grown to the surface over the majority of the outer-shelf reef tract? This is a question worthy of much discussion, because clearly, unknown natural events have interrupted growth during the past 6,000 years.

2

Results of Data Gathering and Mapmaking

Processes II

Seismic-refraction and reflection surveying using dynamite and other explosives to determine deep structures had long been employed in oil exploration. For various environmental reasons, air guns that release bursts of high-pressure air into the water replaced explosives; however, for shallow-water reef and sediment mapping, a different, less powerful, less dangerous technique was needed. In the early 1960s, technicians for Shell Development constructed an electrical sound source called a sparker. Capacitors discharged high-voltage pulses into the water at the tip of a shielded electrical cable.

The discharge vaporizes the water, creating a bubble that, upon collapse, creates a relatively high-frequency sound. The sparking is emitted at rapid rates—usually around one per second. The device is compact and can be used from any skiff large enough to carry a portable electrical generator. The sparking cable is dragged behind the boat along with a hydrophone that receives and amplifies sparker-generated sound. It works much like a fish finder but is much stronger; the sound bounces back microseconds later like an echo from the surfaces of different rock and sediment layers below the seabed. The high-frequency sounds sharpen resolution of the reflections picked up by the hydrophone, and the reflected surfaces are recorded on a moving-paper chart. Today, a laptop computer receives the signals.

In 1964 while working for Shell Development, Paul Enos conducted extensive surveys in the Florida Keys using what was called a mini-sparker. Instead of a single electrode, Shell technicians had devised a sparker with about 62 separate electrodes. This innovation increased resolution by producing smaller bubbles that increased the sound frequency. Paul's work was accomplished from a 6-m (20-ft) skiff. Navigation was by line-of-sight between navigation and other markers as well as by triangulation using a sextant. Global positioning systems (GPSs) were far in the future. Water depth and sediment thickness were ground-truthed by probing with a metal rod, and sediment cores were taken with aluminum tubes pushed and pounded in by divers. Shell Oil Co. eventually released the results of this classic study, which was published as a Geological Society of America Memoir 10 years later (Enos, 1977).

By the late 1970s and during the 1980s, an improved sound source called a boomer came into use. It consisted of a floating frame to which a circular chamber is attached. The chamber has a rubber covering facing downward into the water. An electrical charge activates an internal electromagnet that repulses a metal plate against the rubber cover. The magnetic pulse is like striking a drumhead and delivers a loud thump or "boom" into the water column.

In 1989, the newly established U.S. Geological Survey office in St. Petersburg began a collaborative seismic survey of the Florida Keys National Marine Sanctuary (fig. i.1). The authors worked together with Al Hine and Stan Locker and their students at the University of South Florida St. Petersburg (USFSP) College of Marine Science. The survey was conducted with a boomer, and raw data were recorded on the audio portion of VHS videotape. Maps and profiles resulting from these surveys were published as journal papers (e.g., Lidz et al., 2003) and later digitally compiled with existing and then-new (1997) seismic data and published as a USGS Professional Paper (Lidz et al., 2007).

With the above data-gathering background, we now turn to the actual maps that resulted from these technologies. The maps represent four so-called time-specific stratigraphic[1] layers of data on the Florida shelf. Two of the four maps—bedrock topography and thickness of Holocene accretions, derived primarily from many kilometers of seismic profiles—provided essential data for understanding processes of coral reef evolution along with reef and sediment distribution. The data were hand contoured and digitized (figs. 2.1, 2.2). Because lime sediment has roughly the same seismic velocity (for the return sound signal) as seawater and because of shallow depth and thinness of the sediment cover, the same velocity for sediment, seawater, and depth (i.e., not correcting for any significant difference) was considered sufficiently accurate for our mapping purposes. The two other maps—benthic habitats and sediment composition (both shown later)—were derived primarily from aerial photographs, experience, and field knowledge, and from point counts of sediment grains in thin sections. Those data were also hand contoured and digitized.

To understand modern reef and habitat distribution, it was essential to know the preexisting topography underlying the modern (Holocene) sediments and reefs. Figure 2.1 is a color-enhanced topographic map of that surface (from Lidz, 2000a, and Lidz et al., 2003). Present sea level provides the horizontal datum upon which the bedrock topography is based. In this map, the bedrock surface is the top of the Pleistocene limestone whether exposed or buried beneath Holocene reefs or sediments. The bedrock surface is that which formed as reefs and sediments accreted during the last Pleistocene high sea stand recorded by the corals (at 80 ka, Isotope Substage 5a, fig. 1.1). That sea stand was high enough to flood the shallow shelf but not the island chain or mainland. Minor topographic features may have been slightly enhanced or reduced by erosion when the depositional units were later exposed to air during Isotope Stage 2.

Coring of reefs with a diver-operated rotary coring device (fig. 2.3) and probing with metal rods provided confirmation of seismically determined water depth and Holocene sediment thickness. The data were carefully contoured by hand and later digitized using aerial photography, field experience, and knowledge as guides. This combination of techniques was found to produce far more realistic results along the arcuate reef tract than could be expected from commercial computer-driven con-touring programs. Whereas figure 2.1 depicts the bedrock surface immersed by the Holocene sea, the color-enhanced map in figure 2.2 shows the thickness eventually attained by Holocene sediment and coral accretions (from Lidz, 2000b, and Lidz et al., 2003). The isopach (thickness) map was prepared from seismic-profile data, cores, and probes. As revealed by all existing scientific evidence, when considering the processes (or events) that allowed—or precluded—buildup of Holocene accumu-lations at specific sites or that left their imprint on corals, old shorelines, and exposed land, one can readily visualize the processes by which accretions of the Pleistocene and older times also transpired.[2]

The habitat map (fig. 2.4, from Lidz, 2005, and Lidz et al., 2006) shows the types of seabed settings that overlie the Pleistocene bedrock topography and Holocene reef- and sediment-thickness maps. Habitat types and extents were derived from field ob-servations and prior knowledge augmented by interpreted aerial photomosaics. The habitat map was also hand drawn. Our field observations began in the mid-1970s while at the USGS Fisher Island Field Station in Miami Beach (fig. i.1) and were later continued from the St. Petersburg office. The authors selected the categories[3] for contouring. Coral reefs and debris were divided among living and senile reefs and coral rubble. Presently, many reefs are rapidly shifting into the senile category, and each storm adds to the rubble. The senile-reef classification is the equivalent of what some researchers call hardbottom or live bottom, meaning the hard-rock areas that support live reef-associated organisms but lack an actively accreting reef framework of corals.

Because of persistent water turbidity, many seafloor features were not visible in images of areas south and west of Key West and off the lower Keys. This was espe-cially true of the deeper areas of Hawk Channel farther west of Key West and the area south of the Marquesas-Quicksands sand belt in the Gulf of Mexico, discussed later. Scientists at USFSP have recently surveyed much of that area using sidescan-sonar and swath-bathymetry techniques. Those data (fig. 2.5) indicate the presence of many topographic features of interest in this formerly little-known area.

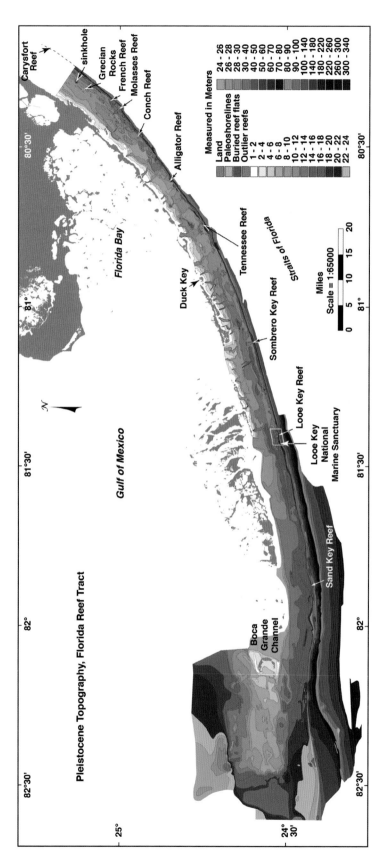

Figure 2.1. Contour map shows regional Pleistocene bedrock surface beneath Holocene and modern reefs and sediments along the south Florida shelf, as interpreted from seismic profiles acquired from the northern Florida Keys to west of The Quicksands. Contour lines can be regarded as shapes and positions of paleoshorelines at different elevations as the Holocene sea encroached on the shallow shelf. Darkest colors = deepest depths below present sea level. Shelf-wide, bedrock elevations are several meters lower to the southwest than northeast, indicating the shelf flooded from southwest to northeast, which is confirmed by northeastward trend of Holocene coral ages. Note generally margin-parallel nature of bedrock trends seaward of the Keys. Contours are in meters. (From Lidz et al., 2003).

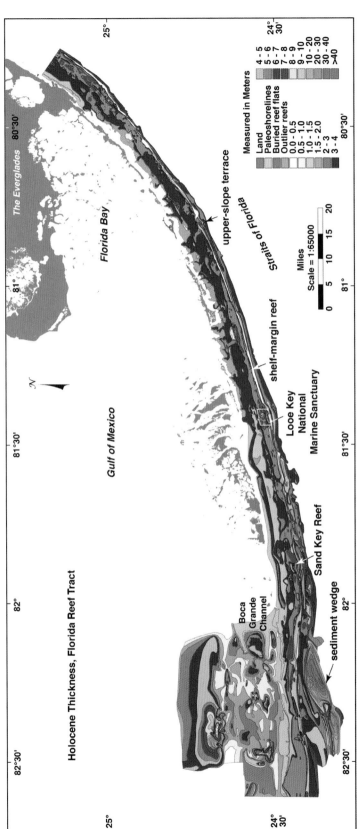

Figure 2.2. Contour map shows regional thickness (isopach) of Holocene and modern reefs and sediments along the south Florida shelf, as interpreted from seismic profiles, probes, and cores acquired from the northern Florida Keys to west of The Quicksands. Darkest colors = thickest accumulations. Reef data are based on drill cores. White area at shelf edge denotes sediment-free seaward side of a massive discontinuous Pleistocene coral reef, which marks the shelf margin at the 30-m (98-ft) depth contour. Sediments are generally ~3–4 m (10–13 ft) thick shelf-wide, indicating offshelf transport in the deeper southwestern area. Contours are in meters. (From Lidz et al., 2003).

Figure 2.3. Tripod and underwater hydraulic drill being used to core Grecian Rocks reef. See figure 3.2A for cross section of reef constructed from results from this core (#4) and similar ones at Grecian Rocks.

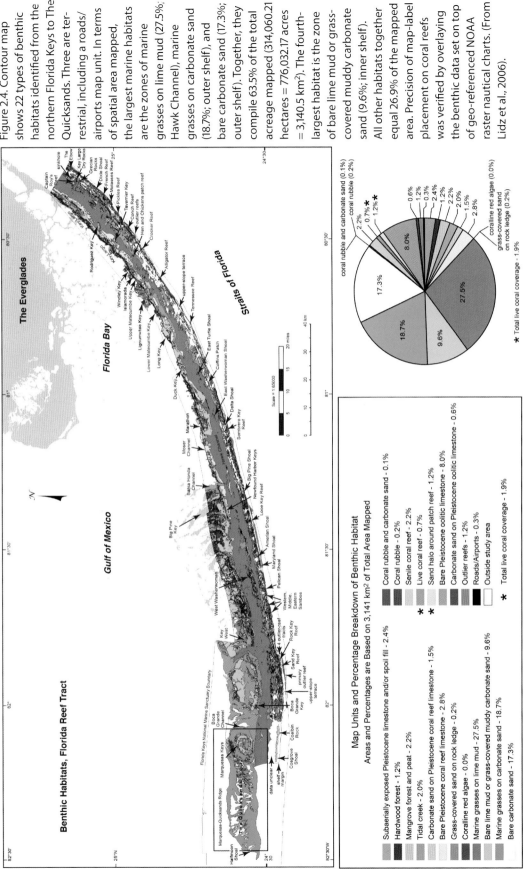

Figure 2.4. Contour map shows 22 types of benthic habitats identified from the northern Florida Keys to The Quicksands. Three are terrestrial, including a roads/airports map unit. In terms of spatial area mapped, the largest marine habitats are the zones of marine grasses on lime mud (27.5%; Hawk Channel), marine grasses on carbonate sand (18.7%; outer shelf), and bare carbonate sand (17.3%; outer shelf). Together, they compile 63.5% of the total acreage mapped (314,060.21 hectares = 776,032.17 acres = 3,140.5 km²). The fourth-largest habitat is the zone of bare lime mud or grass-covered muddy carbonate sand (9.6%; inner shelf). All other habitats together equal 26.9% of the mapped area. Precision of map-label placement on coral reefs was verified by overlaying the benthic data set on top of geo-referenced NOAA raster nautical charts. (From Lidz et al., 2006).

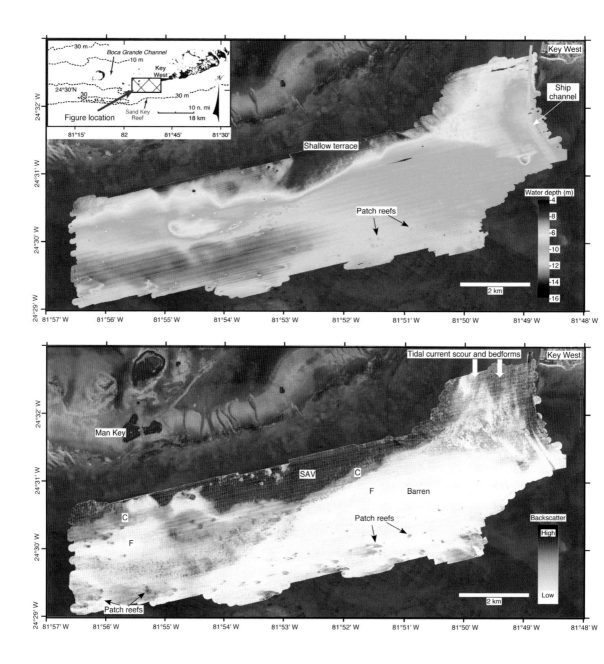

Figure 2.5. Example of seafloor mapping in murky water of Hawk Channel extension south-west of Key West using Teledyne-Benthos C3D sidescan sonar that produces backscatter imagery and bathymetry calculated from the backscatter returns. *Top,* water depth in meters clearly reveals patch reefs and surface irregularities across the nearshore terrace; *bottom,* backscatter imagery provides a better understanding of what is on the bottom, such as veg-etation and differences in sediment type. SAV = submerged aquatic vegetation, C = coarse-grained sediment, F = fine-grained sediment. Vegetation and coarse sand or hardbottoms are more reflective than fine sand. Darker gray is higher backscatter. The effect of bottom currents sorting sediment (streaking) is more apparent in the backscatter image than in the sonar bathymetric image. Unpublished data by Stan Locker.

The fourth geologic map, that of surficial sediment-grain composition, presents the topmost stratigraphic layer along the reef tract. Sediment composition was derived from interpretation of grain origin and grain point counts in thin sections[4] from samples collected shelf-wide in 1989 (fig. 2.6).[5] In essence, three types of skeletal grains dominated sands off the Florida Keys in 1989: *Halimeda*, coral, and mollusk (Lidz and Hallock, 2000). The study contrasted previous, localized analyses by Ginsburg (1956) and Swinchatt (1965), both of which found *Halimeda* to dominate in their studies. The 1989 study showed that in the lower Keys, coral grains far outnumbered other types of grains along the outer shelf and shelf margin. That study directly linked those higher coral-grain percentages to the greater in-situ presence of dead corals and reefs and their postmortem bioerosion[6] and mechanical breakdown. Similar findings have been observed elsewhere in the Caribbean where reefs are severely degraded.

The maps, selected seismic cross sections, and illustrations derived from core data serve as guides for the remainder of the book. They constitute the reader's virtual guide to the evolution and distribution of coral reefs and sediments of the Florida Keys and reef tract, as presently understood. Figure 2.7 shows a sketch of the primary geomorphic features as derived from and confirmed by our studies.

What Is a Reef?

Before proceeding, we should ask an age-old question: "What is a reef?" More specifically, "What is a coral reef?" This seemingly simple question lacks easy answers. Answers remain fuzzy and divided and are confused by maritime history and lore. The traditional sailor's term for reefs includes any topographic feature upon which a vessel may become grounded. Because of this vagueness, areas such as submerged rocks or mountaintops can be called reefs. For example, "Bligh Reef," on which the 274-m (900-ft) Exxon *Valdez* oil tanker grounded on March 25, 1989, is simply a submerged mountaintop—not a true reef. Sandbars have also been called reefs. For the purposes of geology, a more precise definition has long been debated, yet a precise one remains elusive. Geologists lean toward organically created topography, often called bioherms in geologic literature. The key distinction hinges on whether a topographic feature is created by growth of an organism rather than a nonliving object. Thus, there are oyster reefs, sponge reefs, bryozoan reefs, stromatolite reefs, algal reefs, rudistid reefs, and so forth. A sand pile is not a reef. In the Florida Keys, we are concerned mainly with reefs built by corals. On the other hand, there are a few cases where mud banks capped by branching corals have been called reefs.

The more than 1,609-km-long (1,000-mi) Great Barrier Reef off Australia has often served as the model for the Cretaceous reef that encircles the Gulf of Mexico and underlies western Florida (the Sunniland Zone, for example). However, corals of

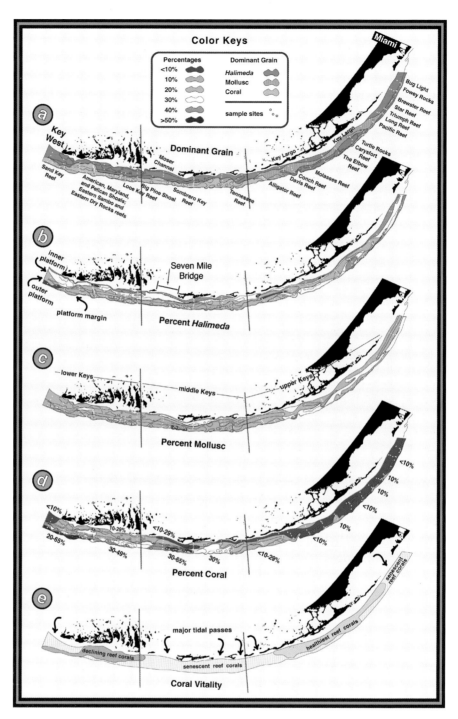

Figure 2.6. Contour maps of dominant grains in sands of the Florida Keys reef tract in 1989. In 1989, *Halimeda* grains prevailed along the entire inner shelf, mollusk grains were dominant along portions of the outer shelf off the upper Keys, and coral grains were dominant along the outer shelf off the middle and lower Keys. The highest proportions of *Halimeda* grains were found in inner-shelf sediments off the northern upper Keys and the lower Keys. Mollusk grains generally composed <10% of skeletal grains off the middle Keys but were ~20% elsewhere. Coral-grain counts were highest off the middle and lower Keys and correlated with reef decline as observed in the field. Vertical red lines show general island divisions into the upper, middle, and lower Keys. Extent and location of Seven Mile Bridge are shown in (b) for scale. (Figure from Lidz, 2000b).

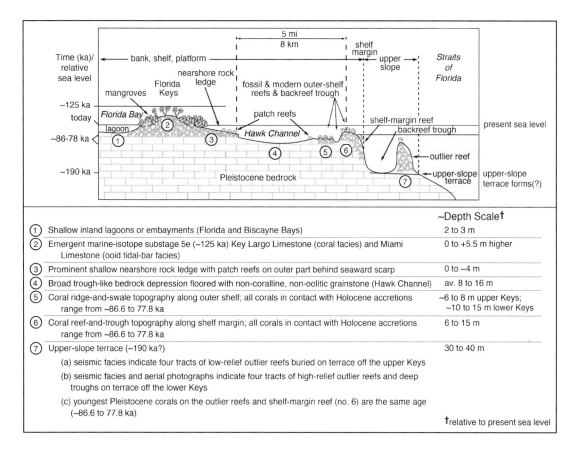

		~Depth Scale†
①	Shallow inland lagoons or embayments (Florida and Biscayne Bays)	2 to 3 m
②	Emergent marine-isotope substage 5e (~125 ka) Key Largo Limestone (coral facies) and Miami Limestone (ooid tidal-bar facies)	0 to +5.5 m higher
③	Prominent shallow nearshore rock ledge with patch reefs on outer part behind seaward scarp	0 to ~4 m
④	Broad trough-like bedrock depression floored with non-coralline, non-oolitic grainstone (Hawk Channel)	av. 8 to 16 m
⑤	Coral ridge-and-swale topography along outer shelf; all corals in contact with Holocene accretions range from ~86.6 to 77.8 ka	~6 to 8 m upper Keys; ~10 to 15 m lower Keys
⑥	Coral reef-and-trough topography along shelf margin; all corals in contact with Holocene accretions range from ~86.6 to 77.8 ka	6 to 15 m
⑦	Upper-slope terrace (~190 ka?)	30 to 40 m
	(a) seismic facies indicate four tracts of low-relief outlier reefs buried on terrace off the upper Keys	
	(b) seismic facies and aerial photographs indicate four tracts of high-relief outlier reefs and deep troughs on terrace off the lower Keys	
	(c) youngest Pleistocene corals on the outlier reefs and shelf-margin reef (no. 6) are the same age (~86.6 to 77.8 ka)	
		†relative to present sea level

the Great Barrier Reef, like the discontinuous outer reefs of Florida, are in places no more than a veneer on top of preexisting topography. And, as in Florida, the thickest buildups lie in the lagoon behind the outer barrier or shelf-edge reefs. The lagoon behind the Great Barrier Reef, however, is many times deeper than that in Florida. The reason modern reefs are not as thick and extensive as in the geologic past is quite simple. Modern coral reefs have been growing only since sea level rose following the last glacial period (Isotope Stage 2), that is, for about 10,000 years, and in Florida only since flooding of the shelf about 7 to 6 ka. Ancient reefs often grew for several million years without frequent interruption by a fluctuating sea level. Although the thick Cretaceous reefs were built mainly by a kind of mollusk called rudistids, some were indeed built by corals similar to those growing today.

Coring of modern coral reefs constantly hammered by waves has often surprisingly revealed a cemented lime-mud–rich matrix. This observation indicates that lime mud filters down into the voids between frame builders as the reef grows. Some lime mud is also thought to precipitate within voids while the reef is growing. These observations based on cores show that the coarse reef sand seen by a diver on a coral reef may not be what is being preserved within the reef framework. This remark should be considered when examining ancient reefs. It should be noted that water

Figure 2.7. Schematic cross section (not to scale) illustrates primary generalized geomorphic features across the south Florida shelf. Sketch is too small to show the White Bank sand bank but its approximate location would be above the circled number 5.

pumped into the core hole to lubricate the drill bit tends to flush out loose lime sediment that has percolated or been precipitated into voids. Cemented lime mud, however, is not flushed away during drilling and can be recovered in the core (fig. 2.8).

The study of reefs and their evolution is fascinating because so many mysteries still remain. As in the past, the best clues will likely come from coring and geologic study of living reefs, as Macintyre (1975) demonstrated in the 1970s with his hydraulic coring device and as shown in figure 2.3.

Although our topic concerns Florida coral reefs and their associated sediment-producing organisms, the remnants of a wide variety of calcium-carbonate–secreting algae and other types of plants and animals often dominate the bulk of what we call reefs. Corals, living or dead, construct the hard surface or foundational framework needed by the attendant void-filling sediment-producing organisms. The Pleistocene Key Largo Limestone that built the middle and upper Florida Keys may be an example of such a composite formation. This so-called reef may have been a carbonate sand bank that included coral patches and isolated head corals. Roughly 50% of this feature is composed of sedimentary grains. The Key Largo Limestone may in fact have been analogous to the present White Bank that will be discussed later.

In the Florida Keys and throughout the Caribbean, the most common noncoral-topography builders are several species of the calcified green alga *Halimeda* (fig. 2.9). *Halimeda* segments or plates the size of oat flakes generally make up much of what we continue to call reefs. Their particles frequently outnumber actual coral fragments. Ginsburg (1956) likened coral reefs to brick buildings, where *Halimeda* and other carbonate-sand grains form the mortar between the bricks. In terms of volume, the amount of mortar often exceeds that of the bricks. Consider a Pleistocene example.

Early analyses by Ginsburg and colleagues at the Shell Development Coral Gables Laboratory showed that coral heads account for only about 49% of the Pleistocene limestone exposed in a quarry now designated the Windley Key Fossil Reef Geological State Park (fig. 2.10A-B). Recent observations have been made that many of the corals are not in their original growth position. The remaining 51% consists mostly of *Halimeda* and other lime-sand particles. Because huge meters-high boulder corals (species of *Montastraea*, *Diploria*, and *Colpophyllia*) stand out so spectacularly in the quarry walls, we tend to call what we see a coral reef. However, close examination in the quarry reveals places where a 10-cm-diameter (4-in.) core would not encounter a single coral. So, was this really a coral reef? Is it possible that what we call the Key Largo Limestone reef was once a sand bank similar to modern White Bank seaward of Key Largo? That is a good discussion for any group visiting the quarry. The discussion of White Bank later may help with the conversation.

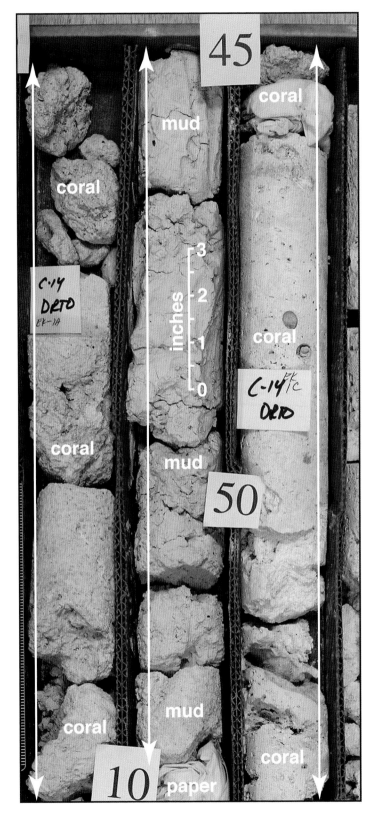

Figure 2.8. Upper portion of an 18-m-long (60-ft) core from a Holocene reef at East Key, Dry Tortugas. Note lime mud fills space between framework corals. Numbers mark intervals in feet (no core recovery between 10 and 45 ft). Labels "C-14 DRTO" indicate core sites sampled for ^{14}C dating. Photo is courtesy of Lauren T. Toth, USGS Research Oceanographer.

Figure 2.9. Some common genera of algal sediment producers from left to right: *Halimeda*, *Penicillus*, *Udotea*, and *Rhipocephalus* (U.S. penny for scale).

Coral Reef Initiation Simplified

High-resolution seismic profiling (Lidz et al., 2007) and underwater diver-operated coring devices (Shinn et al., 1977) have confirmed the importance of underlying topography for coral reef development. Hoffmeister and Ladd (1944) first proposed that antecedent topography exerts a major control on reef development. Purdy (1974) thought such antecedent topography could be created by rainwater dissolution of limestone. For example, if dilute hydrochloric acid is slowly dripped onto a slab of limestone, a depression will form in the middle, leaving elevated ridges around the margin. Does rainwater do the same given time? In the Florida Keys as well as in the Bahamas, preexisting reefs and coastal dunes have been shown to underlie, and thus determine, Holocene reef distribution. Farther back in time, however, and in a broader sense, tectonic deformation and fault scarps ultimately influenced reef distribution in Florida. The antecedent Pleistocene rocks beneath Holocene reefs were probably not exposed to rainwater long enough to create significant shelf-edge topography. The following simplified scenario is considered the most likely explanation for Holocene reef development and distribution in the Florida Keys.

Figure 2.10. *A*, Typical view of Pleistocene Stage-5e limestone in quarry wall at Windley Key Fossil Reef Geological State Park, Florida. A core here would clearly indicate the limestone, known as the Key Largo Limestone, was once a coral reef. Scale in centimeters. *B*, A different view of the quarry wall where obvious corals, except for the small one under the man's hand, are absent. The corals in both photos are species of *Colpophyllia*. A core to the right of the small coral might be interpreted as a fossil sand bank instead of a true coral reef. Thin-section examination would be needed to determine whether the carbonate sand originated as a sand bank or on a coral reef. The accumulation might mistakenly be interpreted as a fossil sand bank like sands of White Bank that are currently forming offshore. The quarry wall in (*B*) is 2 m (6 ft) high.

As the sea slowly encroached on the limestone shelf, drifting coral larvae settled on preexisting, submerged, sediment-free topographic highs and began to grow. At the same time, currents and storms swept carbonate sand produced by breakdown of skeletal coral, *Halimeda,* and other reef-associated organisms into adjacent topographic lows. Generally, coral larvae cannot initiate growth on current-swept, sandy, or muddy sediment. As new corals and sediment-producing organisms flourished, sedimentary grains continued to be shed, filling the lows. Coral growth in the lows was prevented until larger coral fragments were washed in to provide a firm footing for larval settlement. Carbonate sand also accumulated around and between growing corals, but most was swept into the lows. Meanwhile, a rising sea provided increased space for sustained upward coral accretion. Seismic profiling, coring, and radiometric dating reveal a preexisting shelf-edge ridge of 80-ka Pleistocene coral limestone that served as the base for early Holocene coral growth. Because of past fluctuations in Pleistocene sea level, it is likely this reef ridge was in turn controlled by yet older underlying reef growth (Lidz et al., 2008a). Reefs beget reefs; however, cemented Pleistocene and Holocene coastal dunes, oolite, and partially cemented skeletal-sand shoals can also provide topography needed for reef initiation.

Located off the lower Keys about 1.6 km (1 mi) landward of the outer-shelf reef-tract ridge where abundant sediment and coral rubble accumulated are two 80-ka Pleistocene ridges (fig. 2.11A-B). Both were shown by coring to be linear coral reefs. Both lie beneath Holocene coral reefs and reef sediment. In summary, both linear reefs began growing on antecedent coral ridges, and wave action transported loose carbonate sand into the linear topographic low on the landward side of the more seaward ridge. The topographic low, called a backreef trough, parallels the ridges. Coring has demonstrated that the linear low is still being filled with Holocene carbonate sand, similar to what happened during the Pleistocene. A caliche crust provided key evidence. The crust separates uncemented Holocene reef sand in the low from underlying cemented Pleistocene reef sand. Caliche confirms a former period of subaerial exposure between intervals of sand accretion. In this case, reef sand begets reef sand. At the same time, linear coral reefs have grown on topographic highs that straddle either side of the sand-filled low. Cores and coral ages show these Holocene reefs began growth on preexisting Pleistocene reefs. The question becomes, what created the topography on which the Pleistocene reefs developed? Were they cemented beach dunes as are known to occur under nearshore linear reefs off the southeast coast of Florida and in the Bahamas? More-powerful coring or seismic devices will be required to provide an answer.

Coring at other reefs revealed a major reef-limiting factor. Once coral growth reaches the water surface, vertical accretion halts. When this happens, corals are forced to grow laterally or not at all. In all Keys examples studied, reef expansion has

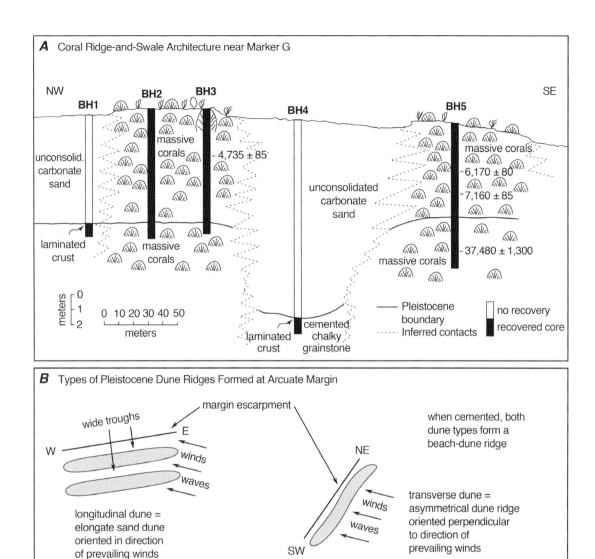

Figure 2.11. *A*, Cross section of core transect across two rock ridges inshore from near Marker G (fig. 3.4A) and southeast of Bahia Honda (BH) in the lower Keys (fig. 1.1). Cores show the ridges are coral reefs separated by a sediment-filled swale. Note lack of a swale/trough on landward NW (left) side of left coral ridge. Also note uncorrected radiocarbon ages of recovered corals. The ^{14}C date of 37,480±1,300 ybp (years before present), obtained on a recrystallized coral, is a result of younger-carbon contamination. Inferred age of bedrock coral sections is between ~86.2 and 77.8 ka (Isotope Substage 5a), which is the age range for the youngest corals elsewhere on the shelf and for those Pleistocene corals in direct contact with Holocene accretions. *B*, Sketch shows processes by which beach dunes can form along an arcuate margin (like the Keys) facing incoming winds and resultant types of dune ridges upon cementation.

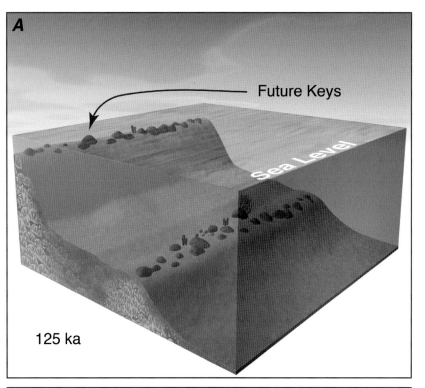

Future Keys

125 ka

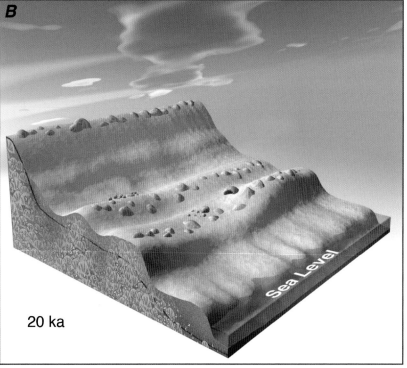

20 ka

Figure 2.12. Changes along the south Florida shelf through time: *A*, At ~125 ka, ancient Ice Age glaciers had melted and caused worldwide ocean levels to rise about 7.6 m (25 ft) above present level. Coral reefs and sandy tidal bars that would become the Florida Keys formed on the shallow shelf. *B*, Glaciers again formed covering continents. Worldwide oceans fell about 122 m (400 ft), exposing the reefs and sediments to air. Beaches created where seawater met land, can be traced around the entire Gulf of Mexico. Trees, grasses, swamps, and soils formed in low areas once covered by seawater. *C*, Glaciers of the last Ice Age once more began to melt, causing oceans to rise again. At ~6 ka, the sea slowly engulfed the old reefs and sediments, and coral larvae again settled on hard substrates. *D*, Glaciers continue to melt today, the seas continue to rise, and coral reefs and sandy reef sediments continue to accumulate. Diagrams courtesy of Florida Institute of Oceanography, images by Betsy Boynton.

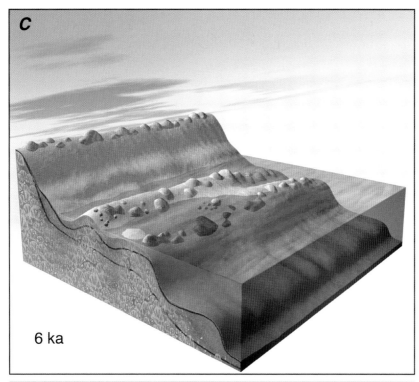

6 ka

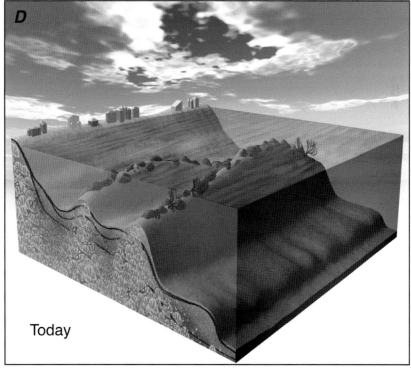

Today

been forced to migrate landward. The process is described as backstepping. Hurricane waves originating in deep water seaward of the reef tract apparently prevent or retard any seaward growth. The same hurricanes break and hurl coral colonies and fragments landward, creating a layer of coral rubble over backreef-sand buildups to form a base for new coral growth, another example of how backstepping can occur. Backstepping can also result as a rising sea changes wave-energy conditions at a reef site, as will be discussed.

The backstepping model, confirmed by coring, is contrary to models exhibited in most textbooks. Textbooks generally depict reefs accreting in a seaward direction, for example the famous Permian (table i.2) El Capitan Reef exposed near Carlsbad, New Mexico. In that case, spectacular outcrops prove the reef did in fact accrete outward into the basin. That model, likely valid for many ancient reefs, has not been observed in the Florida Keys. The controlling factor is twofold: position of sea level and duration at that position. If present sea level in Florida were to remain static for, say, a million or more years, sediment and coral would fill all available marine space on the south Florida shelf, preventing any further upward sediment and reef buildup. Under those conditions, any further reef growth would have to be in a seaward direction and would begin encroaching into the Straits of Florida. This hypothetical model involving static sea level could explain those periods of time when ancient reefs accreted outward into adjacent basins (Shinn et al., 1989).

Block diagrams (fig. 2.12A-D) graphically illustrate the changes in geomorphic features on the south Florida shelf during the last known interglacial period and the last known Ice-Age low sea stand. Note that at around 6 ka, the outer edge of the south Florida platform consisted of a series of Pleistocene islands not unlike those comprising the present Keys. At one point in time while sea level was rising, seawater filled the low swale, now Hawk Channel, behind those islands. This swale was similar to present Florida Bay and was floored with lime mud. Detailed mapping of underlying topography (Lidz and Shinn, 1991) guided by knowledge of the approximate rate of rise of Holocene sea level facilitated delineation of those former islands. With a rising sea, the solid shoreline of each linear island progressively became the foundation for Holocene reef growth. The detailed map of underlying topography (fig. 2.1) was crucial in solving the mystery of why Holocene reefs grew in some places and not in others. Such studies along with radiometric dating of recovered corals also confirmed that reef growth in general began sooner in the lower Keys than in the upper Keys. Why?

Mapping of bedrock topography showed the Pleistocene platform surface slopes to the south and west. A rising sea would first flood the lower-elevation surface off the lower Keys before reaching the higher area off Key Largo. Differences in preexisting topography also explain existence of seemingly anomalous coral areas, such as the patch-reef clusters at Mosquito Bank and elsewhere in the middle of Hawk Channel. Superimposing mapped midchannel patch-reef areas on mapped bedrock topography within the shallow Hawk Channel depression shows that all midchannel patch reefs are aligned along the landward edges of two deeper bedrock depressions behind the shelf-edge outer-reef–tract ridge (Lidz et al., 2006). Many details of modern coral distribution become apparent by study of antecedent topography.

3

Major Geomorphic Topographies

As the sea continued to rise, a mature row of discontinuous reefs developed along the shelf-edge topographic ridge, likely a paleoshoreline discussed earlier. While reef sand was filling the approximately 1.6-km-wide (1-mi) depression behind the outer ridge, corals recruited to the seaward side of the more landward ridge. As the corals grew and backstepped, periodic storms and hurricanes transported carbonate sand landward. The result was a wide, shallow, carbonate-sand bank known locally as White Bank (figs. i.1, 3.1).

White Bank

The best-studied area of White Bank is found off Key Largo, where the sands are as wide as 1.6 km (1 mi) and in places as shallow as 0.6 m (2 ft); however, bank depth off Key Largo is largely about 2 m (6.6 ft) or more, and its surface is ornamented in places by spectacular seagrass meadows and coral-free sand bars and ripples. Although White Bank is predominantly composed of carbonate sand, many clusters of large massive heads of *Montastraea* species occur along its length. Some reach to nearly the sea surface and can be considered hazards to navigation. Push cores and shallow-water seismic profiling showed that White Bank is in places as much as 7.6 m (25 ft) thick (Enos, 1977). White Bank continues northward as a shallow bank beyond the area shown in figure 3.1. South of Key Largo, the bank surface deepens and thickness thins, but the sand feature, although much reduced, remains more or less intact for the length of the reef tract (Lidz et al., 2003, 2006). Its reduction in size and thickness is likely the result of changing shelf-edge geometry, which affects current direction. Off Key Largo, White Bank is oriented perpendicular to the prevailing seas, and tides and currents transport reef sands landward. Farther south where the reef tract curves westward, the bank is oriented mostly parallel to prevailing seas, and sands are transported westward. Coral reefs along the seaward side of White Bank, such

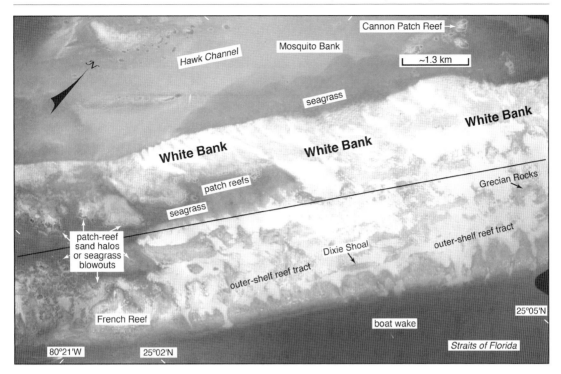

as Grecian Rocks, are the factory areas producing the sediment that built the sand bank. Grecian Rocks originated when the Pleistocene shelf-edge ridge underlying Holocene accretions shielded the area from large waves. Protection from large waves provided low-energy conditions that determined the coral species that would initiate growth of the reef. Detailed coring across the reef (Shinn, 1980) showed growth at Grecian Rocks began with slow-growing massive head corals (fig. 3.2A).

Massive corals such as *Montastraea* species require low-energy conditions and thus prefer either deep water or shallow, protected areas. At Grecian Rocks, the rapidly growing elkhorn coral *Acropora palmata*, which thrives in the surf zone, later replaced the head corals (fig. 3.2A). The flat reef top was created once the elkhorn and other corals grew to sea level. A core drilled behind the reef flat at Grecian Rocks recovered a thin layer of peat that had accumulated on the underlying limestone before corals became established. In summary, the Pleistocene shelf-edge ridge of outer islands that existed around 6 ka when sea level was lower would have protected landward head corals, but as the sea continued to rise, shelter diminished. Fast-growing *A. palmata,* a surf-loving coral, overtook the slower-growing species. Because seismic profiling reveals that essentially no Holocene reef accretion has taken place on the Pleistocene shelf-edge ridge seaward of the area, it is concluded that rising sea level explains the shift from massive quiet-water corals that initiated the reef at Grecian Rocks to the surf-loving branching corals that capped the reef. Concurrent with the rise in sea level and shift in coral species was the landward backstepping of the reef.

Figure 3.1. Aerial photo-mosaic (1975) shows extent of backreef marine-sand belt of White Bank off the upper Keys (area above diagonal black line). Grecian Rocks contributes to White Bank sands. Photos courtesy of Jim Pitts.

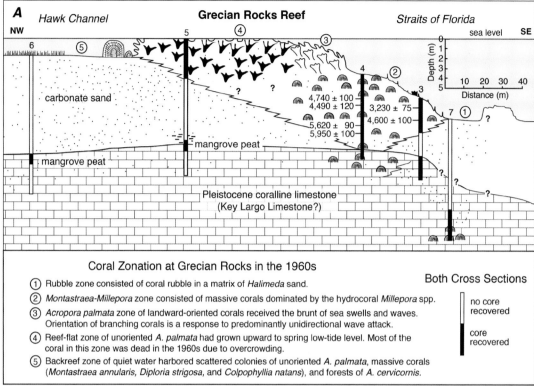

A Hawk Channel **Grecian Rocks Reef** *Straits of Florida*

NW — SE — sea level

6 ⑤ 5 ④ ③ ② 4 3 7 ①

carbonate sand

4,740 ± 100
4,490 ± 120
3,230 ± 75
4,600 ± 100
5,620 ± 90
5,950 ± 100

mangrove peat

mangrove peat

Pleistocene coralline limestone
(Key Largo Limestone?)

Depth (m) 0–5

Distance (m) 10 20 30 40

Coral Zonation at Grecian Rocks in the 1960s

Both Cross Sections

① Rubble zone consisted of coral rubble in a matrix of *Halimeda* sand.

② *Montastraea-Millepora* zone consisted of massive corals dominated by the hydrocoral *Millepora* spp.

③ *Acropora palmata* zone of landward-oriented corals received the brunt of sea swells and waves. Orientation of branching corals is a response to predominantly unidirectional wave attack.

④ Reef-flat zone of unoriented *A. palmata* had grown upward to spring low-tide level. Most of the coral in this zone was dead in the 1960s due to overcrowding.

⑤ Backreef zone of quiet water harbored scattered colonies of unoriented *A. palmata*, massive corals (*Montastraea annularis, Diploria strigosa,* and *Colpophyllia natans*), and forests of *A. cervicornis.*

no core recovered

core recovered

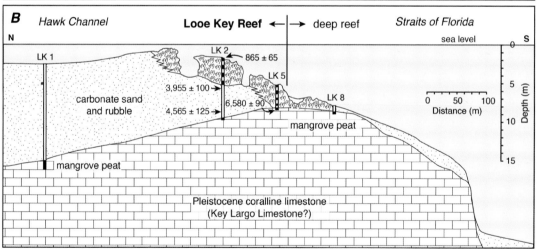

B Hawk Channel **Looe Key Reef** ← | → deep reef *Straits of Florida*

N — S — sea level

LK 1 LK 2 865 ± 65 LK 5 LK 8

3,955 ± 100

carbonate sand and rubble

4,565 ± 125 6,580 ± 90

mangrove peat

mangrove peat

Pleistocene coralline limestone
(Key Largo Limestone?)

Distance (m) 0 50 100

Depth (m) 0 5 10 15

Figure 3.2. *A*, Cross section of Grecian Rocks shows core transect, reef components, and uncorrected radiocarbon ages in years before present (ybp) of corals recovered in cores. Figure 2.3 shows drilling of core #4. Coral zonation in the 1960s showed five distinct zones. Field observations in 2002 found that most corals including hydrocorals were dead. *B*, Cross section across Looe Key Reef (LK) shows the Holocene reef began growth on the crest of a Pleistocene topographic high and backstepped over carbonate sand and rubble. The deepest mangrove peat recovered from the reef tract so far came from a depth of 15 m (49 ft) in core LK 1.

A similar process occurred in lower Keys reefs. At Looe Key Reef, for example, though the coral species (*A. palmata*) remained unchanged, unable to grow out of water, the reef expanded landward, away from waves of the rising sea, and over-topped previously deposited carbonate sand and rubble (fig. 3.2B). Unlike corals on the Great Barrier Reef off Australia, coral growth in Florida was stopped by low-tide level. Corals in the Great Barrier Reef, where the tidal range is much greater, are often exposed to air by at least 30 cm (1 ft) for an hour or more during low tide.

How White Bank widened by landward accretion was revealed after the passage of Hurricane Donna in 1960. In places, as much as 15 m (50 ft) of sand were pushed into Hawk Channel. The sandy sediment was quickly transported and deposited near the angle of repose over the predominantly muddy lime sediment of the channel (Ball et al., 1967). "Angle of repose" is the maximum angle of slope (measured from a horizontal plane) at which loose material will come to rest on a pile of similar material. If one swims over the landward edge of White Bank, an apron of sandy sediment can be observed angling downward from a water depth of around 2.4 m (8 ft) to a depth of 6 m (20 ft) or more. A 15-m-wide (50-ft) strip of marine seagrasses that had carpeted the channel floor was quickly buried during a single hurricane. Assuming one hurricane every 10 years, about 600 such storms could have impacted the reef tract during the 6,000 years it has been submerged.

Hurricane Donna also generated seagrass-free bars of carbonate sand on White Bank. In addition, sand-filled potholes called blowouts were created in seagrass-covered sands landward of several outer reefs such as French Reef (fig. 3.1). The origin of such blowouts had long been the subject of speculation. Subsequent studies by Wanless (1981) showed that, once formed, blowouts tend to migrate laterally. Wave action erodes sand from beneath the seagrass-stabilized seaward side of blowouts while sand simultaneously accretes on the down-current side. In this way, the entire blowout can enlarge or migrate, or both. Hurricanes Donna and Betsy answered many questions.

Hawk Channel

A topographic low or swale marking the landward margin of White Bank forms the seaward side of Hawk Channel. Hawk Channel is a calm well-marked safe-passage body of water that extends from Miami to Key West and beyond. The landward margin of this approximately 3-km-wide (2-mi) topographic low is a sometimes-buried outcrop of Pleistocene bedrock located about 1 km (0.6 mi) seaward of the Keys (fig. 3.3A-B). The outcrop is actually a ledge 0.3–0.6 m high (1–2 ft) thought to be an erosional paleoshoreline. Hawk Channel is partly the result of Pleistocene topography and the sedimentary high created by White Bank. Depth of Hawk Channel in the upper Keys is around 7.6 m (25 ft) but extends to about 12 m (40 ft) in the vicinity of

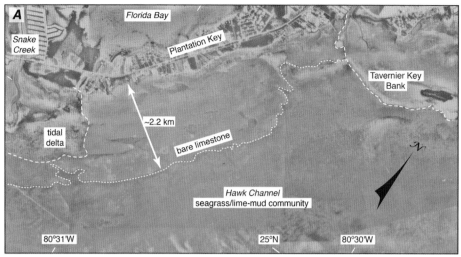

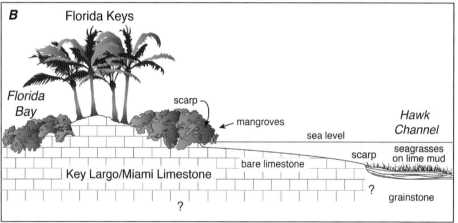

Figure 3.3. *A*, Aerial photomosaic (1975) off Tavernier Key Bank shows a clearly visible ledge of bare limestone (outlined in short dashes) that lines the seaward side of all Florida Keys. A tidal delta at Snake Creek and a muddy bank at Tavernier Key (long dashes) partially cover the ledge off Plantation Key. Photos courtesy of Jim Pitts. *B*, Sketch (not to scale) shows offshore scarp thought to be a paleoshoreline. Scarp forms landward edge of Hawk Channel.

Key West. The southward deepening has long been suspected to indicate differential subsidence of the underlying platform to the southwest. This southwestward deepening is consistent with our contention that flooding and coral growth progressed northward through time. The deep axis of Hawk Channel was likely a topographic low created by the seaward evolution of coral reefs along the outer shelf during the Pleistocene.

Sediment in Hawk Channel is fine grained because of its depth and secluded location leeward of White Bank and the outer-shelf reef tract. The sediment is a mix of lime mud and silt with a variable admixture of molluscan, gastropod, foraminiferal, and pelecypod shells. Except near scattered coral patches, the sediment rarely contains coral. In Dunham (1962) terminology, this sediment is for the most part what will become a wackestone.

More than 90% of Hawk Channel is carpeted with the bladed turtlegrass *Thalassia testudinum* and cylindrical Cuban shoal weed *Halodule wrightii*. Hawk Channel supports the largest area of marine grasses in the entire Florida reef tract (fig. 2.4A-B). Large loggerhead sponges and other smaller species are scattered here and there within the seagrasses. Mud-producing algae such as species of *Penicillus* (fig. 2.9) are relatively abundant, but most of the mud in the channel is probably winnowed from surrounding sand and reef areas and transported there during hurricanes and other storms. Several centimeters (inches) of mud from Florida Bay were deposited in channels throughout the Keys during Hurricane Donna, and the entire reef tract remained turbid for about 30 days following the storm.[1] The sediment is easily stirred into suspension because of the predominately muddy texture. Yachts traveling in Hawk Channel generally create muddy trails that can remain in suspension for several hours.

Many dozens of patch reefs composed mainly of massive corals and gorgonians are found sporadically in Hawk Channel. Such patches often occur in rows parallel to the Keys. Like most Keys reefs, they owe their location to underlying topography (Lidz et al., 2006). Many coral patches lie just below low-tide level and present hazards to navigation. Such areas are mapped and well marked but nonetheless are often the sites of boat groundings and sinkings. Patch reefs are significant features on the Keys reef tract, and further discussion is warranted.

Patch Reefs and Patches of Patch Reefs

The distinction between linear reefs and patch reefs can be a fuzzy one. One can grade into another. Reefs at the edge of the platform and many farther landward are clearly linear. In general, Holocene *Acropora palmata* (fig. 1.13) built the linear ones and initiated growth on linear topographic highs or paleoshorelines. Farther inshore, patch reefs consist of subcircular clumps of massive head corals. Most are found in distinct rows, but some occur haphazardly. In some places, rows of patches join and may appear to be a single linear reef. A characteristic feature of most individual patches is an encircling seagrass-free zone or halo of sandy sediment. At least two different explanations exist for the origin of sand halos. First is foraging by parrotfish and other herbivores that live and hide from predators on the coral patch. They do not range out from the patch very far. The second is that a wave-resistant coral patch intensifies storm currents flowing around the corals, causing erosion and removal of seagrasses. Both processes likely account for the seagrass-free sand halos. Patch reefs

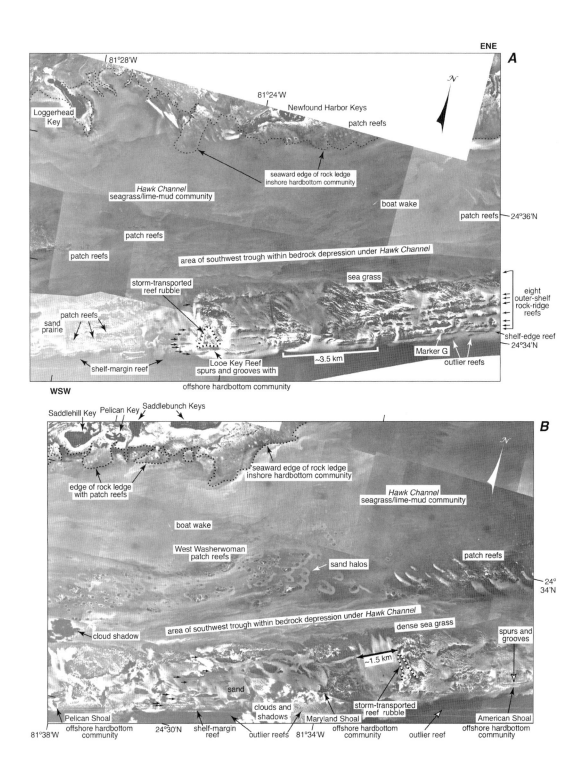

Figure 3.4. Contiguous aerial photomosaics (1975) show seabed features and benthic habitats seaward of (*A*) the Newfound Harbor Keys and (*B*) the Saddlebunch Keys (see fig. 1.1 for locations). Shelf-edge reefs include Looe Key Reef and reefs at American, Maryland, and Pelican Shoals. Note hundreds of patch reefs and their surrounding sand halos. Visible onshore habitats are mangroves. Also note eight linear outer-shelf rock lines (arrows), narrow coral reefs separated by sediment-filled swales such as have been cored near Marker G in (*A*). Photos courtesy of Jim Pitts. Also see figure 2.11A.

and patches of patch reefs in the study area are more abundant off the lower Keys (fig. 3.4A-B) than upper Keys; however, the most abundant patch reefs occur north of the area covered in this book.

Impressive coral patches can be observed landward of certain outer-shelf reefs. Off the upper Keys, large patches occur landward between Carysfort Reef and Hawk Channel (outside the area depicted in the benthic habitat map, fig. 2.4B). Coral patches can be found behind many other coral reefs, some in water as deep as 9 m (30 ft). A large area landward of Long Reef in Biscayne National Park (not shown), north of the area covered in this book, supports literally hundreds of patches at the landward edge of a grassy backreef trough, where depths range from 7.6 to 9 m (25 to 30 ft) (Lidz et al., 2003).

Spectacular patches of giant coral heads form Hen and Chickens Reef located off Snake Creek on the landward margin of Hawk Channel close to shore. Farther south a few kilometers (miles) is another area of massive heads called Cheeca Rocks (fig. 1.6A). Why these patches occur where they do is not well understood but is likely related to underlying topography and lack of sediment cover when coral growth first began. The common factor at both areas is the absence of acroporid corals.

Outer-Shelf or Shelf-Edge Reefs

A series of discontinuous 80-ka Pleistocene reefs forms what could be called a ridge along the shelf margin. Where present, the ridge reefs are backed by backreef troughs either filled or partially filled with Holocene sediment. An upper-slope terrace fronts the toe of the shelf margin in 30 to 40 m (98 to 131 ft) of water. The steep margin face is bare Pleistocene limestone. The Holocene reefs are found along the ridge crest immediately landward of the margin face. Just as coral reef sand and debris contribute to the formation of White Bank, these offshore reefs are the sources of landward-oriented storm-transported rubble fields. Excellent examples include outer-shelf reefs seaward along the platform margin, such as Sombrero Key Reef and Delta Shoal (fig. 3.5). In the case of Looe Key Reef, the rubble field has formed an irregular triangle or "horns" (fig. 3.4A). These reefs are among the thickest in the Keys, and ^{14}C dating of corals in cores shows Holocene growth began along the Pleistocene ridge around 6 ka.

Mudbank Reefs?

Yet another kind of substrate supporting coral growth occurs closer to shore along the landward edge of Hawk Channel. Only two such features are found on the reef tract, Rodriguez Key Bank and Tavernier Key Bank (figs. 1.6A, 3.3A, 3.6A-B). We treat them as reefs because both support a band of branching *Porites divaricata* corals

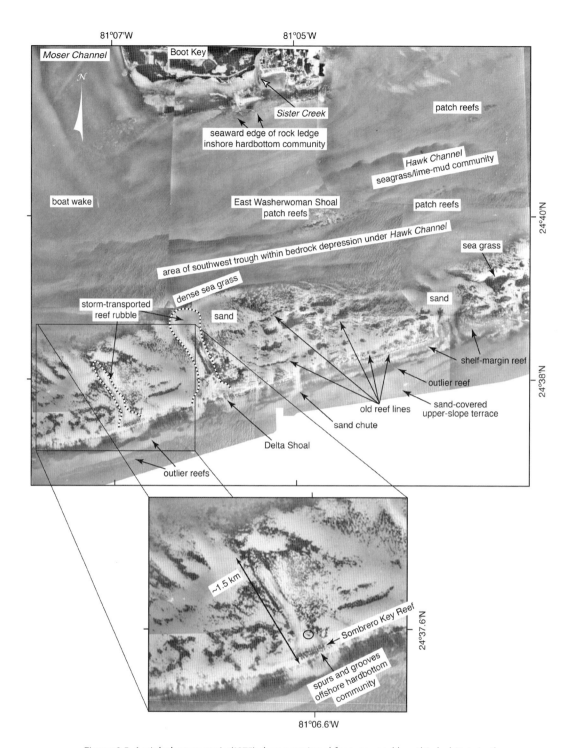

Figure 3.5. Aerial photomosaic (1975) shows regional features and benthic habitats in the area of Boot Key near Moser Channel in the middle Keys. Note elongated areas of storm-transported coral rubble behind Delta Shoal and Sombrero Key Reef (black-and-white dotted lines) and irregularity of skeletal Holocene spurs and grooves at Sombrero Key Reef. Also note numerous areas of sand spillovers or chutes at the shelf-margin reef, caused by low topographic relief, and presence or absence of upper-slope outlier reefs. Sombrero Key Reef Light and its shadow are circled in the inset. Photos courtesy of Jim Pitts.

and coralline algae along their seaward sides. These features have been much studied; geologists once considered them analogous to reeflike oil-bearing buildups in the geologic record. What makes them different? Studies by Turmel and Swanson (1976) showed they are in fact mud banks with a 1.5-m-thick (5-ft) cap of branching corals and coralline algae on their seaward margins. The banks are on average 3.5 m (12 ft) thick and are often awash at spring low tide. Landward of the outer band of branching corals is a wide slightly deeper stretch of highly burrowed mud and sand (what would be a packstone in Dunham terminology). This zone is about 1.5 m (5 ft) thick and overlies more muddy sediment 1.5 to 2.7 m (5 to 6 ft) thick. Burrows made by species of the ghost shrimp *Callianassa* in the upper zone have turned the sediment into something resembling Swiss cheese. Their burrows have likely converted what was lime mud into what will be a packstone when lithified (Shinn, 1968). The lime mud underlying this zone is similar to that which forms mud banks in Florida Bay. Farther inward from the burrowed zone, both Rodriguez and Tavernier Key banks support 1.2-km-long (0.75-mi) mangrove islands for which the banks are named. Geologists have long speculated as to why these two features exist where they do. Were the islands there before the banks formed and did they cause the banks to build on their seaward side, or was it the other way around? Dozens of push cores have been taken on both banks and can be interpreted to support either hypothesis. One feature common to both banks is observed in the cores: the underlying mud looks exactly like muds in cores from Florida Bay mud banks. Carbon-14 dates of shells recovered in cores from the bank bases indicate development began around 5 ka (Turmel and Swanson, 1976). At that time, sea level was lower, and this area probably resembled present Florida Bay. Many hundreds of biologists and geologists have visited these banks as part of organized field trips.

Hurricane Donna revealed yet another unsuspected aspect of lime mud. Whereas offshore coral reefs suffered major damage from the hurricane, Rodriguez and Tavernier Key banks were affected very little. Stiff lime mud is very resistant (fig. 1.9A-B), as was observed during construction of the Flagler Railroad around 1910. Causeways over tidal passes made with riprap to support tracks were washed away in the 1935 Labor Day Hurricane, while those made of lime mud pumped from the bay and allowed to dry resisted erosion. As is observed on mud islands in Florida Bay, the mud stiffens with time and is highly resistant to waves, especially when sun dried.

For reasons not understood, Tavernier Key supports a shelly beach in its center, whereas Rodriguez Key consists entirely of mangrove peat and is often flooded at high tide. Cores taken on Rodriguez Key show the island has been growing toward the southwest while being eroded along its northeast end. In the early days of sailing, it was possible to obtain freshwater from a lens within the beach sand on Tavernier Key (Romans, 1775). It is possible that the mangroves there did not exist in the 1700s. Significantly for modern sailors, both islands provide a calm, protected anchorage on

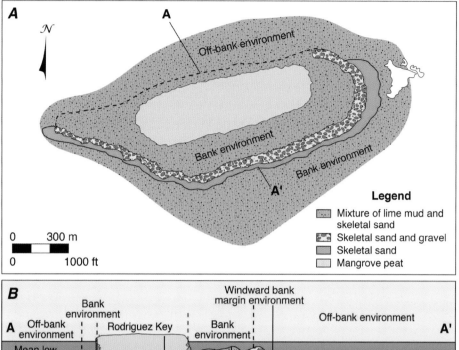

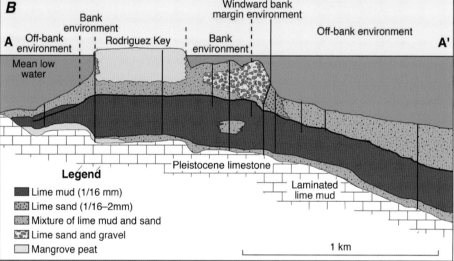

Figure 3.6. *A*, Plan view and (*B*) cross section of components that built Rodriguez Key Bank, upper Florida Keys. Note zones of different surface components in (*A*) and relative extent and thickness of mangrove peat on Rodriguez Key in (*B*), as determined from cores. Also note location of cross section A-A' in (*A*). Redrawn by Betsy Boynton from Turmel and Swanson (1976).

their leeward side. The nearby town of Tavernier on Key Largo may have been named after this Key.

When the senior author was an assistant to Turmel in 1959, one had to swim in from the boat when anchored 30 m (100 ft) seaward of the outer margin of Rodriguez Key Bank. It was too deep to stand. Today one can easily walk in to the shallow bank edge. Clearly, sediment has built up in the 58 years since that work was conducted. A straightforward project waiting for a researcher is to determine how much sediment has accumulated since the original study. Assuming little sea-level rise since 1959, a simple projection of the results of such a study would indicate about when Hawk Channel will fill in and thus no longer be navigable. Now let's turn to processes related to the outer-reef margin.

Spurs and Grooves

Spurs and grooves, sometimes called buttresses, are geomorphic features found on the seaward side of coral reefs worldwide. In Florida, spurs and grooves are the spectacular fingerlike features oriented into oncoming waves on the seaward side of the outermost reefs (figs. 3.7A-B, 3.8) (Shinn, 1963). Swimmers and divers travel great distances to the Keys mainly to swim and dive on spurs and grooves. The value of these features to the dive-charter business and Keys tourist economy is large. Their value to the reefs is also large. Spurs serve as natural breakwaters and reduce damaging impacts of incoming waves. Two distinct kinds of spurs and grooves occur on the reef tract: widely spaced grooves found only on shallow shelf-margin reefs, and narrowly spaced ones in deeper water. The deeper ones formed on fringing reefs when sea level was much lower.

Those reefs that have grown to the surface and are generally marked by lighthouses display the most spectacular spurs and grooves. The spurs with their overhangs and caves range from 4.6 to 15 m (15 to 50 ft) in width and are separated by grooves of about the same width. Tops of the spurs are generally about 3.7 to 4.6 m (12 to 15 ft) below the water surface and stand as high as 3 to 4.6 m (10 to 15 ft) off the bottom.

Some spurs are topped with huge massive coral heads at their deeper seaward ends, whereas intervening grooves are floored with a thin layer of coarse-grained, rippled, white reef sand. All spur-and-groove systems are generally tuned to the direction, height, and frequency of incoming waves. Spurs are little affected by hurricanes, whereas sediment in the grooves is swept landward. Storms thus help preserve spur-and-groove morphology while contributing to shallow backreef sand-and-rubble accumulations.

The origin of spurs has long been debated. Are they constructional in origin, or are they the product of erosion? Whichever the case, once established, sand in the grooves prevents lateral accretion of corals, and those that manage to grow laterally into adjacent grooves are periodically removed by storm waves. Regardless of origin, once formed, spur-and-groove systems tend to perpetuate.

Similar more spectacular systems on Pacific Ocean atolls, where wave action is more severe, tend to be more closely spaced and grooves more deeply eroded. Coral boulders roll back and forth in the grooves and create potholes in response to the larger, more vigorous Pacific waves (Shinn, 2011). Atoll reefs are for the most part intertidal and can thus be considered fringing reefs. The leeward sides of large Pacific

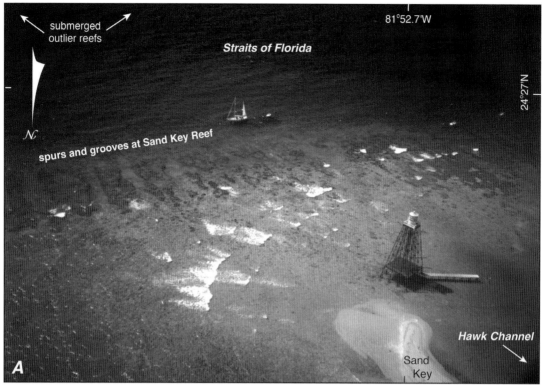

In image A: submerged outlier reefs · Straits of Florida · 81°52.7'W · 24°27'N · spurs and grooves at Sand Key Reef · N · Hawk Channel · Sand Key

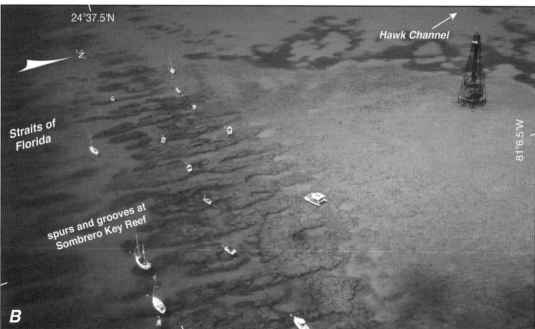

In image B: 24°37.5'N · Hawk Channel · N · Straits of Florida · 81°6.5'W · spurs and grooves at Sombrero Key Reef

Figure 3.7. Spurs and grooves at (A) Sand Key Reef and (B) Sombrero Key Reef offer popular diving sites. The spurs are no longer growing but nonetheless consist of colorful hardbottom communities. Spur-and-groove systems are found on the seaward side of most named reefs throughout the reef tract.

Figure 3.8. Close-up aerial view (1975) highlights multiple spur-and-groove zones at Looe
Key Reef. Formerly known as the "core area" (black dotted lines = core boundary), this area
central to the Looe Key Sanctuary is now called a SPA (Special Protected Area), where diving
is allowed but not fishing. Sand coverage extends seaward and westward, burying each zone
and indicating west-southwest transport directions. Intermediate reef is actually two reef
trends. Deep reef is a section of the shelf-edge reef at the south edge of the sanctuary. Deep
reef here is continuous, extends east and west several hundred meters, and is being buried by
a prominent sand lobe. Rubble "horns" landward of reef zones consist of storm-transported
pebble- to boulder-sized (4–256 mm; 256 mm = ~10 in.) coral debris. Area of reef flat at right is
awash at low tide and was once an island where the crew of the grounded HMS *Looe* lived for
several weeks. Coast Guard Marker 24 is inside southeast corner of core boundary. Courtesy of
Jim Pitts.

atolls, such as Enewetak, generally lack spurs and grooves, and the dropoff into oceanic depths is very steep.

   Today shallow-water spurs in Florida are generally coated with fire coral, also
known as hydrocoral (*Millepora* spp.), that obscures the underlying coral species that
built them. In the early days, dynamite provided a window into the internal structure
of spurs (fig. 3.9) and showed that the fast-growing elkhorn coral *Acropora palmata*
was the main constructor (Shinn, 1963). In recent times, destruction by major ship
groundings and development of underwater coring have further confirmed their
constructional origin. In all Florida cases studied thus far, Holocene elkhorn coral

8 ft
2.4 m

Figure 3.9. Close-up view shows "explosure" at Molasses Reef created by dynamite in 1959 to determine the type of coral that built the reef. Note cross section of skeletal white *Acropora palmata* branches and voids between branches. Most voids were filled with lime mud that was washed away by currents.

was the builder. Unfortunately, because of diseases and coral dieback, the evolutionary biological stages of construction can no longer be observed. Even before culmination of acroporid demise in the early 1980s, few places existed where live elkhorn coral could be observed constructing the spurs. Abundant live *A. palmata* could be found at places like Carysfort Reef, but spurs never matured at Carysfort. The best examples of living spurs were at Grecian Rocks[2] (Shinn, 1963). As recently as the 1950s, the spectacular spurs at Molasses Reef were coated with the fire coral *Millepora complanata*, completely obscuring their internal structure.

Although not presently growing, prime examples of spurs and grooves can be observed at The Elbow, Molasses Reef, French Reef, Sombrero Key Reef, and Looe Key Reef (figs. 3.7A-B, 3.8). Although Carysfort is a mature Holocene reef 13 m (43 ft) thick, spurs are poorly developed there. Their absence is related to rising sea level and a broad, shallow, shielding, seaward outlier reef that will be discussed later.

Core borings in grooves for emplacement of mooring-buoy anchors revealed that the rock underlying the thin layer of sand is Pleistocene limestone. The surface of this limestone is generally coated with red-brown caliche, thus proving the surface was formerly dry land. The caliche coating also shows that, unlike grooves on Pacific atolls, negligible wave-induced erosion has occurred. In addition, the coral spurs also initiated growth directly on the caliche-coated Pleistocene surface. The short height (3–3.7 m, 10–12 ft) of such spurs indicates the small amount of coral accumulation that has occurred on Florida reefs during the past 6,000 years. Given known coral-growth rates, spurs should have reached the sea surface in less than 1,000 years. In most cases, elkhorn spurs most likely did reach sea level, at which time vertical growth ended. Sea level continued to rise, but dead spurs for various reasons were not recolonized and did not resume growth. Another question of, why not?

Spurs growing on a level caliche-coated surface demonstrate that preexisting topography did not determine their distribution and spacing. Carbon-14 dating of coral at the base of a spur near its deeper seaward end at Looe Key Reef shows that development there began around 6,580 ± 90 ybp (years before present), or ~6.5 ka (fig. 3.2B). Again the amount of upward growth is minimal considering time and water space available for growth.

A characteristic feature of reefs with mature spurs and grooves is that the spurs and grooves lie seaward of a reef flat composed mainly of coral rubble and reef sand. In some places, the rubble has formed an island. Historically such islands were larger, which explains why the word "key" is part of the name of a reef. Looe Key Reef is a good example. The sand spit exposed there at low tide is the remnant of a once larger island (see fig. 3.8 and caption). The 280 survivors of the HMS *Looe* that grounded there in 1744 endured on that island until they eventually sailed away on the launches recovered from their wrecked ship.

Hurricane Donna in 1960 removed Little Molasses Island north of Molasses Reef. Though the 1.5-m-high (5-ft) coral-rubble island was washed away, it soon reformed (Ball et al., 1967). When the senior author first camped on the island in the 1950s, it supported small bushes. Six years after its destruction, the island had enlarged sufficiently in size that a truck-mounted drill rig could be driven off a landing craft onto the sand and rubble. The rig cored 13.7 m (45 ft) of Holocene coral rubble before penetrating Pleistocene limestone (Perkins, 1977). The rubble and sand at Sand Key Reef 13 km (8 mi) off Key West was equally thick. In 1827, an 18-m-tall (60-ft) brick lighthouse tower was built on Sand Key Reef along with a separate building as home for the lightkeeper and family. Unfortunately, all were washed away by a hurricane in 1846. The island today is much smaller (fig. 3.7A), and a 40-m-tall (132-ft) iron lighthouse was constructed there in 1853.[3] The authors once rescued six Cuban refugees who had been dropped off on this sand-spit island by a small boat from Havana.

Fringing-Reef Spurs

A distinctly different spur-and-groove system begins seaward of the main reefs in around 20 m (65 ft) of water and extends seaward to a depth of around 30 m (100 ft).

These spurs and grooves are evenly spaced and much narrower than those in shallow water. They are around 3 m (10 ft) wide and seldom more than 0.3 to 0.6 m (1 to 2 ft) high. The grooves are roughly the same width, and the reef sand is less than 0.3 m (1 ft) thick. They resemble those on the seaward side of Pacific atolls. The grooves are assumed to be erosional and are interpreted as typical of those that form on fringing reefs and limestone shorelines the world over (Shinn, 2011). Good examples can be seen on the seaward side of what we now call outlier reefs (Lidz et al., 1991).

Backreef Ledges and Paleo-Sea Levels

Several platform-margin reefs display spectacular overhanging ledges facing shoreward. Such ledges, several hundred feet long, range from 6 to 8 m (20 to 25 ft) below sea level and from 1.5 to 2.4 m (5 to 8 ft) in height. The ledges rise up from a flat sand-covered surface. Such ledges occur south of Alligator Reef, and at Davis and Crocker reefs. These ledge sites provide a unique window into coral reef evolution.

Excavation at the base of an extensive overhanging ledge south of Alligator Reef revealed organic peat overlying a caliche layer (fig. 3.10A,C). The peat here was dated at 8 ka, and a core through the top of the ledge revealed it had also been built by *A. palmata*. It too had begun growing on caliche (Robbin, 1981). Carbon dating of coral recovered in the core revealed an age of approximately 6 ka. Beneath a few centimeters (inches) of reef sand at the base of the Davis Reef ledge, a well-developed caliche layer coats the underlying Pleistocene limestone (fig. 3.10B,D). A bulk ^{14}C-age revealed a date of approximately 14 ka; thus, this was dry land around 14 ka. The overhanging ledge only a few feet away was constructed by *A. palmata* grown directly on the caliche layer.

In-situ *A. palmata* constructed both ledges described here and can be observed by scuba diving along the ledge bases. The uniformly flat surfaces of the ledge tops indicate the position of sea level when the corals stopped growing. Like spurs described earlier, upward growth was undoubtedly halted by a static sea level. Why growth did not resume later is not understood but most likely indicated a change in water quality or temperature unfavorable for later growth. The eventual opening of channels into Florida Bay and release of inimical bay water as sea level rose could have terminated *A. palmata* growth. Future research should be directed toward these ledges. They may have implications for development of paleoshorelines elsewhere along the reef tract. It should be noted that because of the rapid growth rate of *A. palmata* (~10 cm/yr, 3.9 in./yr), the in-situ colonies that built the ledges and overlying reef flats could have easily done so in less than 50 years given optimal conditions.

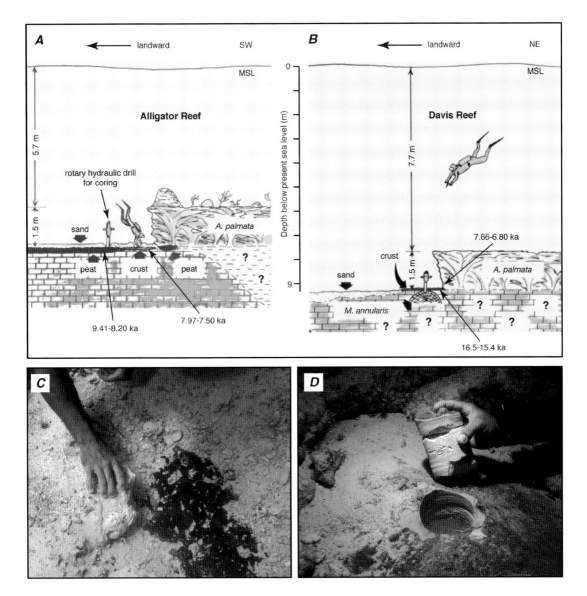

Figure 3.10. *A-B*, Sketches drawn by Harold Hudson show locations of calcrete relative to present sea level and proximity to coral reefs at Alligator and Davis reefs in the upper Keys. Ages have been converted from calendar ybp to ka for brevity. Corrected ^{14}C-age range (7.97–7.50 ka) for Alligator peat indicates reef growth before ~7.5 ka would not have been possible because of subaerial conditions. An older age range (9.41–8.20 ka) for underlying crust indicates the peat was forming in a moisture-laden depression at Alligator Reef site. Corrected ^{14}C-age range (16.5–15.4 ka) for top 2 cm (0.8 in.) of crust at deeper Davis Reef site is consistent with submergence of deeper site during crust formation at Alligator Reef. The corrected age range for coral at the Davis site (7.66–6.80 ka) indicates that corals at the Alligator site are probably younger than 6.8 ka as a result of later flooding of higher-elevation bedrock at Alligator Reef. MSL = mean sea level. *C*, Mangrove peat overlies calcrete (caliche) under ~30 cm (~12 in.) of coarse sand at a site within ~5 m (16 ft) of shoreward margin of Alligator Reef. *D*, Cored crust caps Pleistocene *Montastraea annularis,* under diver's thumb, at Davis Reef. Base of dead *Acropora palmata* reef is at upper left. Uncorrected date of crust was 14 ka.

Outlier Reefs

Seaward of several shelf-margin reefs, especially off Sand Key Reef and Carysfort Reef (locations on fig. 3.11A-B), discontinuous linear topographic/geomorphic features occur that have been termed "outlier reefs" in the scientific literature (fig. 3.12A-D; Lidz et al., 1991; Lidz, 2006). Dates on cored outlier corals indicate the reefs formed at different times during different periods of Pleistocene marine transgressions and regressions.[4] These previously anonymous features were called outliers because they grew on an upper-slope terrace seaward of the main chain of shelf-margin reefs and are separated from the platform margin by deep backreef troughs. Four such outlier tracts are found off both lower and upper Keys, but usually only the largest is obvious and appears on official charts. None of them are named (named reefs are Holocene accretions).

Off the lower Keys (Sand Key Reef), where the terrace is 1 km (~0.6 mi) wide and about 40 m (130 ft) deep, the largest of the four outlier tracts grew to ~30 m (~100 ft) of relief as seen on seismic profiles. The ridge-shaped outlier fronts an unfilled trough (fig. 3.12C-D). Off the upper Keys (Carysfort, The Elbow), the four tracts are low in relief and buried in Holocene sands (fig. 3.12A-B). The main outlier seaward of Carysfort Reef is as shallow as 6.7 m (22 ft) and separated from Carysfort by an 18-m-deep (60-ft) trough partially filled with sediment (fig. 3.12A). Although the Carysfort outlier is 22 m (~72 ft) thick, it did not attain as much relief as the primary outlier off Sand Key Reef. When sea level was much lower, the outlier reefs would have provided protection to shelf-edge reefs from large waves, especially at Carysfort Reef because of proximity of the Carysfort outlier to what was then a shelf-edge reef on which Carysfort grew. That Carysfort Reef was protected from high-energy waves during its growth is supported by the lack of a mature spur-and-groove system.

In all cases, the discontinuous outlier tracts lie parallel to the main shelf-edge reef tract and occupy an upper-slope terrace that varies in width and a few meters in depth along the platform margin. Their vestiges occur sporadically along the terrace (figs. 3.13A-B, 3.14A-B, 3.15A-B, 3.16A-B; Lidz et al., 1997, 2007). Judging from coral ages and corresponding coral depths below present sea level throughout the reef tract, the terrace is believed to be about 190 ka (MIS 7) (Lidz et al., 2008a).

Radiometric dates on cored corals from the largest outlier off Sand Key Reef indicate that the reef grew during three late Pleistocene eustatic cycles: during isotope-Substages 5c (coral date of 106.5 ka), 5b (coral dates of 94.4–90.4 ka), and 5a (coral dates of 84.5–80.0 ka) times. Whether the sea was in transgressive or regressive phases during those depositional periods is not known. The tops of the outliers formed during Substage 5a between 86.6 and 80 ka (Ludwig et al., 1996), the same age and time as Pleistocene corals now directly under Holocene accretions on the shelf. Judging from ages and depths of the outlier corals relative to present sea level,

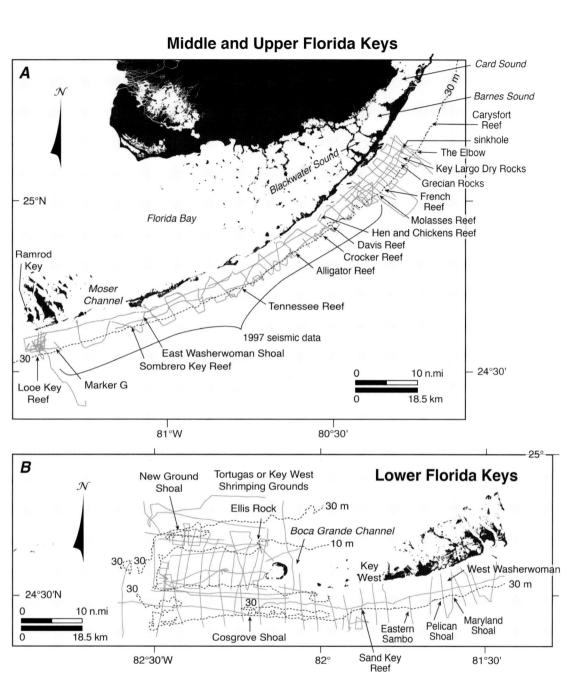

Figure 3.11. *A-B*, Contiguous index maps show locations of major Holocene reefs and USGS geophysical surveys (gray lines) off the Florida Keys. Note location of sinkhole at northeast end of seismic lines. Sinkhole is more than 55 m (180 ft) deep, as determined from probing and seismic profiles, and is filled with sediment. The sinkhole is the only one known to date to occur on the south Florida shelf.

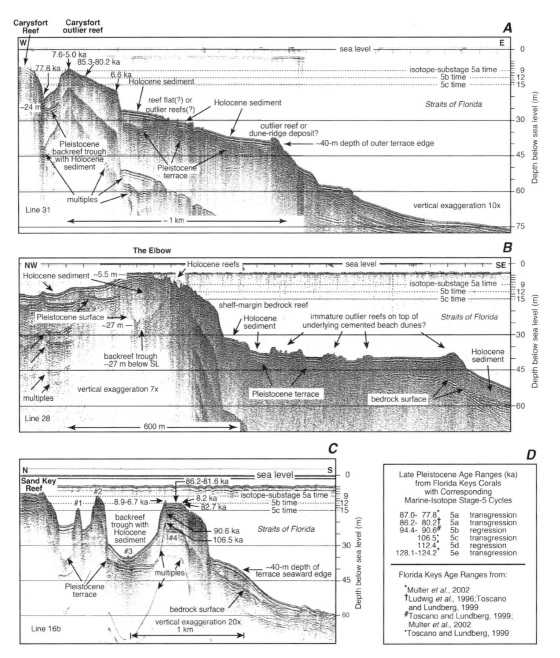

Figure 3.12. *A*, Seismic profile (perpendicular to reef tract) across Holocene Carysfort Reef (upper Keys, see fig. 3.11A for location) and its seaward Pleistocene outlier reef shows geomorphology of Pleistocene and Holocene buildups in the area. Note coral dates. Youngest Pleistocene date on Florida reef tract is 77.8 ka, obtained on a coral recovered from below Carysfort Reef. Different dates and isotope substages indicate different periods of growth during different positions of Pleistocene sea level. Multiples = seismic energy that has been reflected off a surface more than once. *B*, Seismic profile (oblique to reef tract) crossed Holocene patch reefs and The Elbow (upper Keys, fig. 3.11A). Note infilled backreef trough behind The Elbow and four possible immature outlier reefs on seaward upper-slope terrace. Different isotope substages indicate different periods of accretion during different positions of Pleistocene sea level. *C*, Seismic profile (perpendicular to reef tract) crossed Holocene Sand Key Reef (lower Keys, fig. 1.1) and seaward outlier reefs. Note mostly empty backreef troughs and tall seismic relief of outliers. Also note coral ages and correlative (positions of sea level during) isotope-substage depositional times. *D*, Legend correlates interpreted age ranges from Florida Keys corals with Marine Isotope Stage-5 transgressional and regressional cycles.

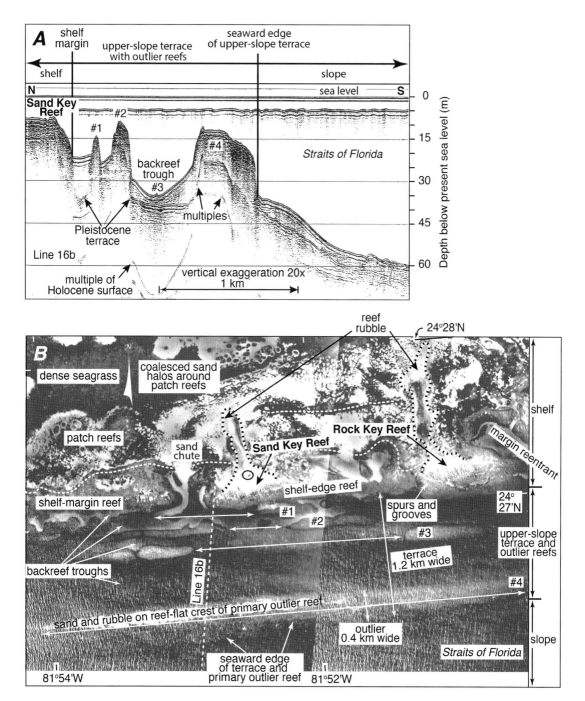

Figure 3.13. *A*, Seismic profile, same as shown with coral ages in figure 3.12C, demonstrates that seismic surveys are only as accurate as the course of the vessel over the bottom. In this case, outlier reef tract #3 was not conclusively known until (*B*) aerial photographs (courtesy of Jim Pitts, 1975) revealed its existence as well as the discontinuity of the four outlier tracts. Sand Key Light is inside the black circle. Dashed white lines parallel to shelf margin mark old reef lines. White dashed seismic Line 16b corresponds to location of profile shown in (*A*) and figure 3.12C.

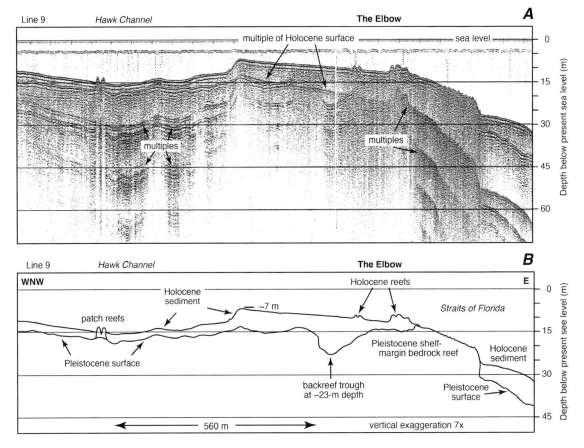

Figure 3.14. *A-B*, Seismic profile (tangential to the shelf edge, with interpretation) shows broad extent and thickness of Holocene sediment accretion behind the Pleistocene shelf-margin reef beneath The Elbow.

late Pleistocene sea levels apparently reached highstand maxima of –15 m at Substage 5c, –14 to –10 m at Substage 5b, and –9 m at Substage 5a (–49, –46 to –33, and –30 ft, respectively) (Lidz et al., 2008a). Sea level at 5c and 5b times was not high enough to immerse the Substage-5e (125 ka) Florida Keys, or the mainland. The last Pleistocene sea to flood the shelf occurred during Substage-5a time.

Coring shows the Sand Key Reef outliers are composed mainly of massive head corals. Fossil acroporid corals were not detected in the cores. These outliers would have been tall, elongate, coral-rock islands when the shelf-margin reefs they sheltered began growing toward the end of Isotope Stage-2 time. A closely spaced spur-and-groove system typical of fringing reefs can be found on the seaward side of all the outliers throughout the Keys. They most likely formed toward the end of Stage 2 and represent an old shoreline.

A mystery not yet solved by coring is what initiated and imposed the linear geometry on these unique features. We speculate they began growth on cemented coastal beach dunes similar to reef nuclei in the Bahamas (see fig. 2.11B). For technical and mechanical reasons, our coring of the outliers terminated in Substage-5a strata (fig. 1.1). For one, the drill site is in open water ~12 m (~40 ft) deep with strong currents, creating a difficult setting for divers operating drilling equipment. Equipment was also insufficient to core deep enough (>30 m, or >100 ft) to test the sand dune hypothesis. Such dunes composed of oolite do occur in much deeper water, as shown earlier (Locker et al., 1996). For that reason and the fact that Holocene coastal dunes are known to underlie linear nearshore reefs off northern Miami Beach (Shinn, 1976),

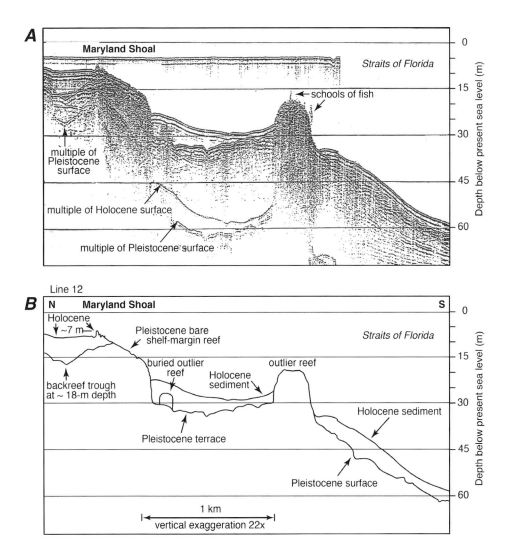

Figure 3.15. *A-B*, Seismic profile (perpendicular to shelf edge, with interpretation) crossed Maryland Shoal (lower Keys shelf edge, see fig. 3.4B for location). Note the smaller of two outlier reefs is buried in Holocene sediments sloughed seaward from the shelf margin. Also note infilled backreef trough and bare limestone face of shelf-margin reef.

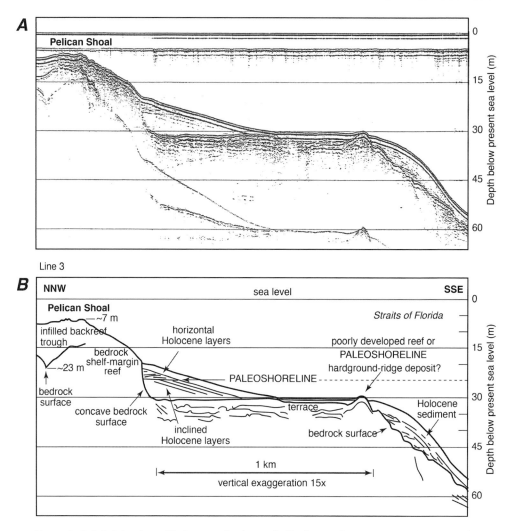

Figure 3.16. *A-B*, Seismic profile (perpendicular to shelf edge, with interpretation) across Pelican Shoal (lower Keys, fig. 3.4B) shows 0.6-mi-wide (1-km) upper-slope terrace and what may be interpreted as the nucleus (cemented beach-dune ridge?) of an incipient outlier reef at seaward edge of terrace. Note difference in sediment layers at foot of shelf margin. Older (bottom) diagonal layers have been truncated by a stillstand in sea level followed by deposition of younger horizontal layers at the paleoshoreline. Also note infilled backreef trough and bare limestone face of shelf-margin reef.

we propose that similar coastal dunes formed, became cemented, and thus provided the solid nuclei for coral growth and creation of the outlier reefs.

A few head corals and dense thickets of dead Holocene staghorn coral (*A. cervicornis*) blanketed large areas of the primary Sand Key Reef outlier. A length of nylon anchor line ~30 m (~100 ft) long was recovered from beneath a thick carpet of dead staghorn branches. Death had occurred recently, most likely around 1980. The ongoing plague of regional coral demise is discussed later in the book.

Only a thin coating of mostly dead Holocene coral was found on the main outlier reefs cored elsewhere in the Keys. That situation changes farther north in the vicinity of Miami. With the exception of the four buried low-relief tracts on the upper-slope terrace, as the outlier tracts closer to the present shelf margin are traced northward, they increasingly support thicker accumulations of Holocene coral. Coring showed that the Holocene coral outlier in 14 m (45 ft) of water seaward of Fowey Rocks Reef, a shelf-edge reef (fig. i.1), is 15 m (49 ft) thick. That now dead reef had been constructed almost entirely of *Acropora palmata*. Carbon-14 dates from core intervals are provided in the sea-level curve in figure 1.5. The core bottomed in quartz sand thought to be a coastal dune after penetrating the 15 m (49 ft) of Holocene coral.

Once again, coring raised more mysteries. In most places where *A. palmata* had constructed Holocene reefs, the underlying Pleistocene was composed of massive head corals. How was it that a species that prefers nonturbulent conditions grew in places adjacent to deep water that at times, when the position of sea level was changing, must have been turbulent and where *A. palmata* would be expected? Massive head corals presently do not generally grow in shallow, wave-battered zones at the margin of a platform. Similar observations were made at a site in 15 m (50 ft) of water adjacent to the *Aquarius* underwater-habitat site seaward of Conch Reef (fig. 2.1A), where a core revealed Pleistocene reef rock built entirely by head corals. The core contained no warm-temperature shallow-water acroporid species, and no acroporids presently grow there. The major surprise was that the overlying Holocene was only about 15 cm (6 in.) thick and underlain by caliche! With changing sea levels, this site was certainly within the turbulent surf zone at various times in the recent past. Why was no *A. palmata* present? Was the Pleistocene water too cold for this temperature-sensitive species? Much remains to be learned.

A Giant Sediment Trap

A major serendipitous discovery in the early 1990s unveiled a huge infilled sinkhole on the Florida reef tract in 5 to 7 m (16 to 23 ft) of water off Key Largo (fig. 3.11A). The sinkhole, which measures 600 m (~1,970 ft) in diameter, is located at 25°08'37.30"N and 80°17'50.34"W (fig. 3.17A). The sinkhole is almost impossible to see from a boat and thus had not been discovered until observed on aerial photography. (It was initially thought to be a water spot on the photographic negative!) Once we had seen it, we thought it might be a filled meteor-impact crater; however, coring, probing, and seismic profiling confirmed it was in fact a sediment-filled sinkhole. Several hundred such features are well known in the Bahamas, where they are commonly called

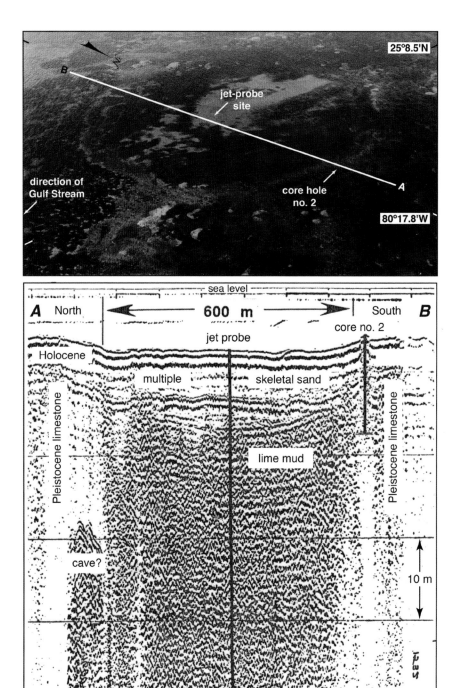

Figure 3.17. *A*, Oblique aerial view toward the southwest of 600-m-diameter (~1,970-ft) infilled sinkhole shows locations of limestone core and jet probe. Core 2 outside sinkhole and another out of view showed no evidence of fracturing as from an impact. Reef corals ring the feature, but none grow within the sediment-filled hole. Dark areas are marine grasses. *B*, Seismic profile along transect A-A' shows lime-sand fill overlying gas-filled lime mud. Jet probe penetrated 55 m (180 ft) of mud, but total thickness is unknown. Mud collected from bottom of jet probe provided a bulk radiocarbon date of ~5.7 ka. Feature to left is interpreted to be a cave filled with lime mud.

blueholes. After two cores from surrounding limestone showed a lack of fractures, a meteoric-impact origin was ruled out.

Initial rotary coring in the center of the feature recovered a few meters of carbonate sand underlain by gooey lime mud. High-resolution seismic profiles showed a seismic signature characteristic of gas-filled sediment (fig. 3.17B). Subsequent push coring with 7.6-cm-diameter (3-in.) aluminum tubes to a depth of 32 m (105 ft) recovered featureless aragonitic lime mud below the overlying sand layer. The limited power of our underwater coring equipment prevented coring to deeper depths. To penetrate deeper, we used water pipes in 3-m (10-ft) lengths to jet probe to a depth of 54.5 m (179 ft). Resistance of the probe caused by stiff lime mud prevented further penetration. We estimate, based on known depths of similar Bahamian blueholes, that the sinkhole is at least 120 m (~395 ft) deep. Radiocarbon dating of mud recovered from the bottom of the jet probe 54.3 m (~178 ft) below the seabed revealed a bulk age of ~5.7 ka. The youngest mud age just below the carbonate sand layer was ~3.3 ka. The dates indicated that the sinkhole had acted as a giant sediment trap that began filling rapidly once the area was flooded by rising sea level around 6 ka. Results of our study were published in the *Journal of Coastal Research* (Shinn et al., 1996). To our knowledge, this is the only sinkhole of this type found in the Florida Keys, and it is much larger in diameter than those common to the Bahamas. A larger, more powerful barge-mounted drill will be required to core and sample its entire depth and sediment thickness. A core to rock bottom may provide a Holocene record of climate and sea-level changes from when the sinkhole was likely a freshwater-filled sinkhole similar to the cenotes commonly found on the Yucatan Peninsula of Mexico.

We now switch to a different type of environment—one that is most likely controlled by geomorphic orientation and greater exposure to deleterious Gulf of Mexico waters.

4

Western Terminus of the Reef Tract

The general arcuate-orientation theme of Hawk Channel, White Bank, and outer-shelf coral reefs parallels the curvature of the shelf margin, trending north-south off the upper Keys, transitioning to northeast-southwest, and finally to east-west in the lower Keys. The reef system continues westward some 64 km (40 mi) beyond Key West and terminates at Halfmoon Shoal at the west edge of a shallow Pleistocene limestone platform (figs. i.1, 4.1A-B). The platform begins just west of the natural channel deepened to create the Key West Harbor. The environmental setting on the platform is different from that off the Keys. Less than 2.5 m (8 ft) deep at its eastern end, this part of the platform supports mangrove islands, mud banks, and several low Pleistocene islands. For the most part, sediment is thin, and abundant sponges grow directly on the bedrock. This area has long supported a commercial sponging industry. Because of reduced Holocene sedimentation and because the main focus of our mapping efforts was Holocene reefs and sediments, we did not map this area in detail.

Geomorphologies of The Quicksands

The westernmost Pleistocene island on the west end of the platform is Boca Grande Key (fig. 4.1A-B). Just west of that island, the surface deepens to form a natural 8-km-wide (5-mi) 3- to 5.5-m-deep (10- to 18-ft) current-swept channel called Boca Grande Channel. Little Holocene sediment has accumulated in the central axis of this north-south channel because of strong reversing tidal currents; however, sediment thickness increases rapidly at the west edge of the channel. The area of thickened sediment is the beginning of a large swath of Holocene lime sand, composed mainly of *Halimeda opuntia* plates, that covers much of the western end of what we informally call the Marquesas-Quicksands ridge (fig. i.1). The belt of carbonate sand, officially known as The Quicksands, is ornamented with large submarine sand waves that mi-grate back and forth (north and south) on the ridge. A ring of islands named the Marquesas Keys marks the east end of the ridge (fig. 4.1B). The 4.8-km-diameter (3-mi) island ring is not rock but loose *Halimeda* sand populated by mangroves, palm trees, seagrasses, and various tropical hardwoods.

Within the island ring, soft burrowed lime mud and silt support marine grasses and large patches of living *Halimeda*. In a study conducted there, Hudson (1985) found that a single *Halimeda* plate geometrically multiplied into 3,000 plates in just 125 days! *Halimeda* is clearly a prolific sediment producer, not just in the Marquesas Keys and The Quicksands but also throughout the Caribbean. Probing with a metal rod showed the central mud-and-silt mixture is approximately 5 m (~16 ft) thick. A tidal channel extending around the northwest end of the main island into the island ring and often exposed at low tide is ~4 m (13 ft) deep (fig. 4.1A). Bare Pleistocene limestone lines the channel bottom.

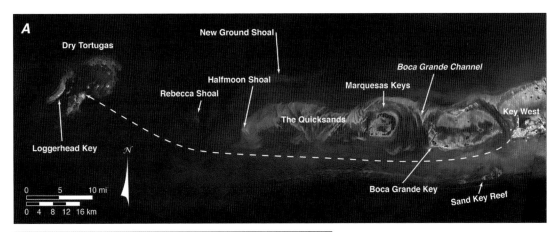

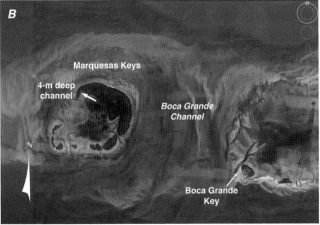

Figure 4.1. *A*, Enhanced Thematic Map-per Plus image, acquired in February 2000 from the Landsat-7 satellite, shows key geographic features and proposed fieldtrip route (dashed line) between Key West and islands of the Dry Tortugas. *B*, Google Earth image from Landsat-7 satellite (2014) shows details of the Marquesas Keys and Boca Grande Channel. A 4-m-deep (13-ft) channel curves around the northwest end of the main island, hugging that part of the island until it breaks away and enters the central island ring (arrow).

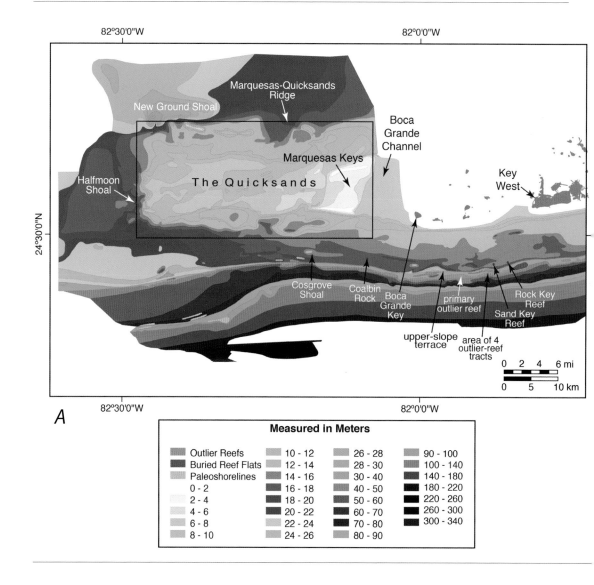

A

Bedrock-topography maps reveal that sands and fine-grained sediment of the Marquesas Keys have accumulated in an east-west–oriented topographic depression between two linear ridges (fig. 4.2A-B). The north ridge consists of a series of oolite crests interpreted to be late Pleistocene beach deposits (Shinn et al., 1977, 1990). Oriented cores show the oolite is cross-bedded, and most beds dip toward the north. Dip directions and keystone vugs[1] (Dunham, 1970) indicate the oolite ridges are beachrock and thus represent approximate sea level at the time of formation. Water depth is as shallow as 1.2 m (4 ft). About 40 km (25 mi) to the east, the same belt of paleo-oolite beachrock crops out along the north side of several of the north-south–trending lower Keys. There it presently lies within the intertidal zone. This exposed

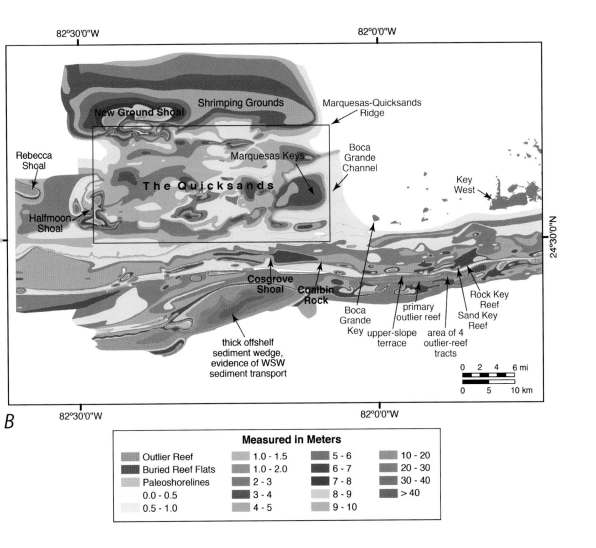

Figure 4.2. *A*, Colored contour map shows bedrock-surface topography beneath Holocene reefs and sediments on the Marquesas-Quicksands (M-Q) ridge and its adjacent areas. Note narrow shallow rock ridges (linear white features) on north and south sides of the Marquesas Keys and westward orientation of the main ridge, which is surrounded by deeper bedrock. New Ground Shoal is coralline limestone. Limestone on the M-Q ridge is oolite. (Modified from Lidz et al., 2003). *B*, Colored contour map shows thickness of Holocene reefs and sediments in the same area. Both maps are derived from data from high-resolution seismic profiles, standard marine charts, aerial photographs, and field knowledge. (Modified from Lidz et al., 2003).

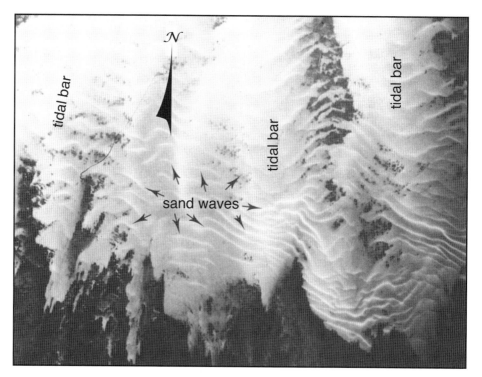

Figure 4.3. Pre-1990 aerial photomosaic shows a portion of The Quicksands west of the Marquesas Keys. Note sand waves are perpendicular to tidal bars. Dark areas between sandbars are seagrasses and *Halimeda*. Sands on the ridge are composed primarily of fragmented plates of *Halimeda opuntia* (Hudson, 1985).

east-west paleo-beachrock is considered the same age as the paleo-beach beneath 1.2 m (4 ft) of water on the north side of the Marquesas Keys. Because beach deposits are good indicators of the position of sea level at the time of deposition, we view this relation as additional evidence of differential platform subsidence. The mystery is, when did differential subsidence occur?

In contrast to the Pleistocene limestone ridge along the north side of the Marquesas Keys, the ridge off the south side of the Marquesas is deeper and lies 1.5 to 2.4 m (5 to 8 ft) below the water surface. The south ridge consists of burrowed pelleted-oolite grainstone (Shinn et al., 1990) with no beach bedding or bedding of any kind. Scattered patches of massive *Montastraea* heads occur where the burrowed limestone is free of sediment. This limestone is very different from the paleo-beach ridges off the north side of the islands. The Marquesas Keys shelter corals that grow on the south ridge from the influence of cold Gulf of Mexico waters during winter months.

Extending westward is the vast 32-km-long (20-mi) area of shifting *Halimeda* sands called The Quicksands, so noted on navigational charts. The sands form wide north-south–trending bars ornamented with 1.8-m-high (6-ft) east-west–trending mega-ripples that can be described as sand waves or subaqueous dunes (fig. 4.3). Our

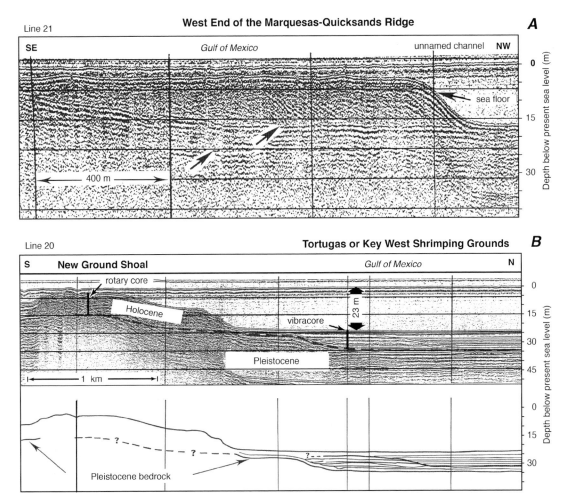

Figure 4.4. *A*, East-west section of high-resolution seismic profile at west end of The Quicksands shows westward accretion of Holocene sediments. Accretionary westward-dipping laminations are composed mainly of *Halimeda* sand. Note contact between uncemented sand and underlying oolitic bedrock (top diagonal arrow). *B*, North-south section of seismic profile and interpretation show changes in the seafloor surface and composition from across the reef at New Ground Shoal into the muddy carbonate sediments of what is known as the Shrimping Grounds. Note location of a 6-m-long (20-ft) sediment vibracore and site of a 7.5-m-long (25-ft) rotary core through the Holocene coral reef overlying Pleistocene coralline limestone.

simple measurements with implanted rods indicate some of these sand waves migrate back and forth about 9 m (30 ft) with each tidal change. Is this why the area is called The Quicksands, or is it because many ships have disappeared[2] beneath the sands?

In places, sand-wave crests reach the sea surface during low tides. High-resolution seismic surveys show sand thicknesses of 3 to 6 m (10 to 20 ft) culminating at Halfmoon Shoal to the west, where sediment is as much as 12 m thick (40 ft). Seismic profiles (fig. 4.4A) demonstrate net accretion is to the west, and sand spills into a

>15-m-deep (50-ft) unnamed channel separating The Quicksands from the Tortugas Bank, another 32 km (20 mi) farther west (figs. i.1, i.2, 4.1A). Little is known about the lithology and biota of this deeper area. The water is always turbid, and the bottom is never visible from the surface.

One large topographic feature called Rebecca Shoal occurs in this channel (figs. 4.1A, 4.2A-B). The top of the shoal is only a few feet deep and once supported a building on iron poles that served as a lighthouse. To our knowledge, no coring or seismic surveys have been conducted on Rebecca Shoal.

Rimming the north edge of the Marquesas-Quicksands ridge and paralleling The Quicksands belt is an east-west–trending series of dead Holocene coral reefs. The western terminus of the reef trend 12 km (7.5 mi) north of Halfmoon Shoal is a shallow reef called New Ground Shoal (figs. i.1, 4.1A, 4.2B). Coring there revealed a 7.6-m-thick (25-ft) reef built by massive head corals. The core contained no *Acropora*, and none were living there when cores were drilled in 1982. The seismic profile across New Ground Shoal (fig. 4.4B) shows the core location. Cores at a similar dead-reef accretion called Ellis Rock farther east along this trend revealed the same composition. *Acropora* had never played a role in the building of these reefs. Despite this fact, the area is now regarded as "critical habitat" for acroporids.[3] We deployed a thermograph at New Ground Shoal to investigate the probable cause of lack of acroporid corals. During winter, water temperature fell below that required for acroporid growth (13°C, 55°F) (Shinn, 1984). The water is apparently too cold for recruitment, let alone continued growth, of acroporid corals. The area lies in the path of cold Loop Currents that sporadically flow out from the northern Gulf of Mexico during winter months.

South of The Quicksands sand belt, the water that covers the continuation of the bedrock depression of Hawk Channel is deeper and more turbid than off the upper and middle Keys. The few patch reefs that do occur there can seldom be seen from the sea surface. The bottom half of the Spanish-treasure galleon *Nuestra Señora de Atocha* lies on a topographic feature composed of oolite in 14.6 m (48 ft) of water. The ship hit the oolite in the swale of enormous waves during a hurricane in 1622. Carrying tons of silver, the bottom broke loose and sank, while the upper part of the ship and its cannons were strewn a great distance across The Quicksands. After many years of searching, treasure salver Mel Fisher and his crew recovered around 36,287 kg (40 tons) of silver bars, gold coins and chains, precious stones, and other treasure. Remains of the sister ship *Margarita* lie nearby. The *Margarita* had been found in shallower water, and its treasure was salvaged shortly after it sank. The topographic feature on which the *Atocha* sank is oolitic limestone.[4]

Farther offshore lies the east-west continuation of the shelf-margin reef tract. The accumulation of coral and reef sand is thin and for the most part consists of exposed Pleistocene limestone. This area is subject to the same cold Gulf waters that retard

Figure 4.5. Oblique aerial view (2004) of Fort Jefferson, an asymmetrical hexagonal structure, on Garden Key shows a sand bridge with Bush Key. A strong storm built the bridge. Hurricane Katrina reopened the former channel in 2005.

growth at New Ground Shoal and Ellis Rock. Three oil-test wells were drilled along this trend, as noted earlier (fig. i.5). Farther offshore in 80 m (263 ft) of water lie the 14-ka coastal dunes and oolite described by Locker et al. (1996).

The continuation of the Hawk Channel depression separates the outer shelf-margin reef tract from The Quicksands belt. The open water is never clear enough to see the seabed from the surface. Parts of this deeper turbid area of the Hawk Channel extension, known as West Channel, have been mapped using sidescan-sonar and swath-bathymetry surveys (fig. 2.5A-B, unpublished data by Stan Locker).

The Dry Tortugas

We now skip past the 32-km-wide (20-mi) channel that separates Halfmoon Shoal from what Spanish explorer Ponce de León named "Las Tortugas" to the west, where many turtles thrived when he discovered the atoll-shaped ring of islands and reefs in 1513 (fig. i.2). "Dry" was later added to "Tortugas" because freshwater is nonexistent there. The Tortugas is best known for Fort Jefferson, the largest brick fort (16 million bricks) in the United States (fig. 4.5). Construction of Fort Jefferson on Garden Key began in 1846, but it was never fully completed and its cannons never fired at an enemy. Its main use was as a coaling station and refuge for U.S. Navy ships. The battleship USS *Maine* took on coal at the fort before exploding in Havana Harbor in 1898. Though the cause was never determined, the incident propelled the United States into the Spanish-American War. During the earlier Civil War (1861–65), the fort also served as a prison. The entire world knows that Dr. Samuel A. Mudd, the man who set the broken leg of John Wilkes Booth after he assassinated President Abraham Lincoln, was imprisoned there. At one time, close to 2,000 soldiers were stationed at this remote fort. The moat around the fort served for sewage disposal during military occupancy. Seawater was allowed to enter the moat at high tide and flush out the foul water. Soldiers favored the windward side of the fort. Fort Jefferson, named after President Thomas Jefferson, was designated a National Monument by Executive Order of President Franklin D. Roosevelt in 1935. The entire Dry Tortugas and Fort Jefferson did not formerly become a National Park until 1992.

As mentioned, the Carnegie Institution Research Laboratory was built on Loggerhead Key in 1904 and received its first scientists in 1905. Mapping of the area by Alexander Agassiz in 1881 and the so-called black-water events of 1878 and 1889 helped Alfred Goldsboro Mayer convince Carnegie Institution officials to fund establishment of the remote laboratory. During its existence on Loggerhead Key from 1905 to 1939, many coral-growth experiments were conducted within the moat around the fort, which at the time was essentially abandoned. Most coral work, however, was conducted in the shallows west of Loggerhead Key on what is now called African Reef (fig. i.2). The 46-m-tall (150-ft) brick lighthouse on Loggerhead Key was completed in 1858, so the lighthouse keepers were the only other people on the island when the Carnegie Laboratory was established. The original, much smaller Tortugas lighthouse had been located on Garden Key and was still in operation when the fort was being constructed around it in 1846.

Origin of the 27 × 12 km (17 × 7.5 mi) atoll-shaped islands of the Tortugas has long been discussed, and much has been published on the topic. Is it a true atoll, or does it simply resemble an atoll? That elusive question stimulated studies by Alexander Agassiz, who produced the first habitat map of the Tortugas in 1881 (published in Agassiz, 1883). Later, in 1914, T. Wayland Vaughan, who became a principal researcher at the lab, called both the Marquesas Keys and Dry Tortugas atolls; however, he realized islands of the Marquesas were composed of sediment and those of the Tortugas were primarily coral. He also knew they were not related to Darwinian-type atolls in the Pacific, whose origins have volcanic roots. Results of studies at the Loggerhead laboratory (which included such diverse subjects as ant behavior) were published mainly in the *Proceedings of the Carnegie Institution* in Washington, DC. Ironically, a laboratory scientist concluded that fine-grained lime mud precipitated under the influence of bacteria (Drew, 1914). For a while, lime mud was known as drewite, and its origin remained controversial for many years. Now, more than 100 years later, many scientists have come to the same conclusion as to its source.

Acroporids at Dry Tortugas are sporadically killed by frigid water imported on Loop Currents that bring it southward from the northern Gulf of Mexico. Frigid water most likely explains why acroporids were not detected in any of the cores drilled there. Acroporid corals contributed little to the development of Tortugas Bank. The main builders were massive species of *Montastraea*, *Diploria*, and *Colpophyllia*. A well-recorded incident, known as the black-water event, impacted the Tortugas in 1878. That event, most likely what we today call a red tide, wiped out most of the Florida west-coast fisheries and killed enormous numbers of fish and corals at the Tortugas (Mayer, 1903). Publication of that event in the journal *Science* helped Mayer establish the Carnegie Laboratory at Loggerhead Key.

Beachrock

Although intertidal beachrock has been identified in coral reef areas the world over for more than 100 years, for reasons not known, the Dry Tortugas is the only place in Florida where beachrock is presently forming (figs. i.3A-B, 4.6A-B). Early workers at the Carnegie Laboratory documented its presence, and one experiment indicated it could form in one year. The first paper published by Ginsburg (1953) was on the formation of beachrock at Loggerhead Key. What is known is that when the chemistry is right, tidal pumping of seawater within beach sand stimulates precipitation of acicular aragonite needles around sand grains. This process produces layers of very hard limestone. Such limestone always contains keystone vugs. Beachrock is clearly visible along the west side of Loggerhead Key in aerial photographs (fig. i.3A-B). The seawater pipe embedded in beachrock (fig. 4.6B) was installed in uncemented beach sand around 1905. Beachrock now below sea level can also be observed around Hospital Key in Dry Tortugas (fig. i.2; Freode and Shinn, 2012).

Modern Research at Tortugas

In 1976, the National Park Service hosted a number of researchers at Garden Key, including members of the USGS Fisher Island Station. While other scientists conducted biological surveys, the USGS team was busy doing what the original Carnegie researchers could not—coring surrounding reefs and islands to unravel their recorded geologic history. Using their newly developed underwater coring device, the USGS team drilled six reef cores, including one at the site of the former Carnegie Laboratory. Major findings were (a) the thickest, most prominent Holocene reef throughout

Figure 4.6. *A*, View (2004) on west side of Loggerhead Key shows beachrock offshore from site of the Carnegie Laboratory dock. In 1976, beach sand extended seaward out to beachrock in the background. The sand has been washed away. *B*, Iron pipe (sewer or seawater intake?) emplaced in sand in 1905 is now encased in beachrock and attests to rapidity of beachrock formation.

the Florida Keys, called Southeast Reef, is located southeast of Fort Jefferson, (b) the reef is 17 m (55 ft) thick, and (c) the reef had prograded seaward, unlike other reefs drilled on the Florida reef tract (fig. 4.7A-B). Another significant finding was that, also unlike Florida reefs, the Holocene elkhorn coral *Acropora palmata* had never been a significant reef builder. During the study, only one small living colony was discovered. Many small patches of *A. palmata* have been located since then. On the other hand, living staghorn coral *A. cervicornis* was abundant and formed extensive thickets in most shallow areas. The vast majority present in 1976 was killed during a severe cold front the following year (Davis, 1982). The storm was severe enough to bring frost and snow flurries to Miami and Homestead. In addition, small colonies of *A. palmata* transplanted by the USGS team from the Keys the previous year had also succumbed. Staghorn coral is currently making a comeback, but based on history revealed in our cores, its future in the Dry Tortugas is considered tenuous. Staghorn coral in fact seldom contributes significantly to reef making, even in the Florida Keys. In spite of its fast-growing and conspicuous delicate beauty, staghorn coral can be likened to a weed compared with other reef-building species.

In 2004, the USGS returned to Garden Key and installed water-monitoring wells at two sites within the fort and at five sites in the moat outside the fort. The purpose was to determine groundwater-flow direction and to monitor for fecal bacteria. Photographs of the drilling team and equipment can be viewed at http://soundwaves.

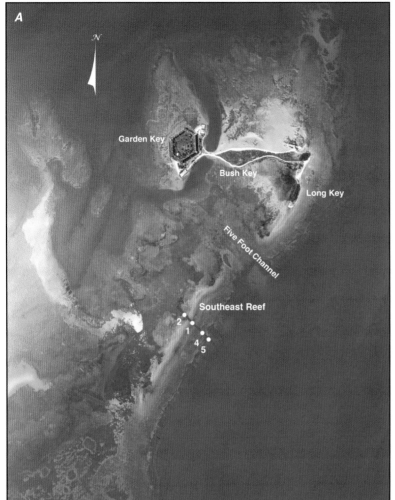

Figure 4.7. *A*, Aerial photo shows Fort Jefferson on Garden Key, sand bridge from Garden Key to Bush Key, and core transect (A-A') across Southeast Reef, Dry Tortugas. Core #3 was too short to show any geology. *B*, Geologic cross section derived from the four core holes drilled on transect A-A' at Southeast Reef shows configuration of underlying Pleistocene coralline limestone and ^{14}C-age dates of massive corals recovered in core #2. *Acropora* species were not encountered in the cores. Core #3 from a live coral head was obtained to study growth bands.

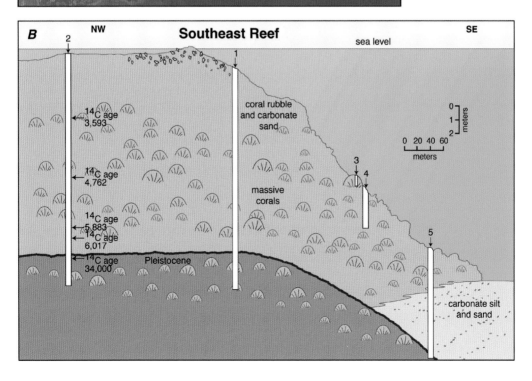

usgs.gov/2004/10/fieldwork2.html (accessed 11/17/16) and http://soundwaves.usgs. gov/2005/09/fieldwork4.html (accessed 11/17/16).

Researchers from the USGS and University of South Florida St. Petersburg (USFSP) have recently conducted lidar, sidescan, and swath surveys of most of the area, and various groups continue to conduct research there. Currently, USGS geologists and biologists are carrying out long-term growth-rate studies as proxies for past climatic conditions. Researchers are also transplanting corals to determine recent and future climate effects in comparison with similar experiments being conducted in the Florida Keys. Researchers today have many modern tools not available to the pre-scuba-diving scientists at the Carnegie Laboratory. Let's now change gears and discuss other factors related to the Florida Keys, its coral reefs, and its people.

5

Coral Health,
or Lack Thereof

That corals at the Dry Tortugas and throughout the Florida Keys have been experiencing ongoing decline since the late 1970s and early 1980s is well documented. Concern about overall coral health throughout the Caribbean has been increasing over the past two decades. Keys reefs are no exception. Determining the causes of coral demise is difficult, because most mortality occurred pretty much in parallel with population expansion in the Keys and in the Caribbean and because so many factors may be involved. Many explanations for the demise of coral reefs in the Keys have been proposed since the 1970s, when reef degradation also began to be noticed throughout the Caribbean. Clearly, the mortality of corals parallels rampant increases in tourism and immigration, which, in the Keys, swelled with the introduction of abundant freshwater, mosquito control, new bridges, and, significantly, the lack of major hurricane strikes since 1965. All the above, and more, have made the Keys a destination not only for tourists but also for retirees as well as working people wishing to get away from it all. With that said, we next briefly discuss a few possible causes that have been proposed for long-term demise of Keys coral reefs.

Sewage

With such explosive population growth on a narrow chain of islands, one must wonder, where does the poop go? Obviously, it goes into the ocean but not before taking a circuitous detour through porous and permeable Substage-5e Key Largo Limestone. Septic tanks and shallow-disposal wells, the original traditional solutions, were replaced by more septic tanks and even more shallow-disposal wells. Fortunately, all are currently being switched to centralized sewerage systems.[1] The traditional solution for hotels and restaurants was to inject treated sewage into wells ranging from 17 to 27 m (55 to 90 ft) in depth. Before injection, solids were removed, and liquids were aerated, treated with chlorine tablets in so-called "package plants," and gravity did the rest. The centralized systems will also treat the sewage; nonetheless, it too will still go into the ground but at much deeper depths, around >853 m (2,800 ft). Similar to treatment systems in Miami and other parts of Florida, the effluent will be pumped into a cavernous Eocene (table i.2) limestone called the Boulder Zone. Pumping is required because the lower part of the Boulder Zone south of Miami is artesian to the extent that enough pressure is present to push water >3.7 m (12 or more feet) above sea level. During drilling of the oil test well (by Gulf Oil Co.) on the reef tract south of the Marquesas Keys in 1960, water rose 19 m (75 ft) to the rig floor when the Boulder Zone was encountered (personal observation of David Folger, on-site geologist present on the rig). Until the system is completed, toilet waste continues to be shunted into around 30,000 septic tanks, some with and some without drain fields, and into 1,000 shallow-disposal wells. And where does the limestone receiving the treatment-system wastes lead? To the Atlantic Ocean.

The water table, either brackish or saline, ranges from just under your feet to as much as 3.7 m (12 ft) deep. Because the water table is so close to the surface, the bottoms of septic tanks were often knocked out to prevent them from floating out of the ground during high tides. For the most part, the table is less than 1.8 m (6 ft) deep, and it is just the first stop on the way of natural groundwater flow to the ocean. Many dozens of USGS water-monitoring wells injected with fluorescein dye, radioactive tracers, and bacterial and viral tracers have shown that net groundwater flow is toward the coral reefs—and it moves fast. Because of Keys limestone porosity and permeability, flow is on average 2.5 m (8 ft) per day (Reich et al., 2002).

Why is flow toward the Atlantic? The Yucatan Current pushes water into the Gulf of Mexico with such force that it raises water level in the Gulf and in Florida Bay. Direct measurements show that water level on the bay side of Key Largo, where tidal fluctuation is very little is, on average, about 15 cm higher (6 in.) than mean sea level on the Atlantic side. During low tide on the Atlantic side, the level on the bay side can be as much as 1 m (3 ft) higher. Water always finds its way downhill, and the porous and permeable limestone simply slows the flow rate but does not stop it. During high tide on the Atlantic side, water level is lower in the bay but for too short a time to overcome net seaward flow. On average, the level remains higher in the bay (fig. 5.1A-B), so except during storms, net flow remains toward the Atlantic and the coral reefs (Shinn et al., 1994; Reich et al., 2002). This intermittent-flow rate toward the Atlantic transports not only nutrients from fertilizers and sewage but also bacteria, viruses, pharmaceuticals, animal wastes, asphalt runoff, and other contaminants (Paul et al., 1995a,b, and many others). These observations support the contention of some researchers who believe sewage is the most significant cause of coral disease and mortality (Porter et al., 1999). Others believe boat exhausts, oils, and other human-related factors are contributory agents. Precht and Miller (2007) have provided an extensive review of suspected causes of reef degradation. Nevertheless, whatever

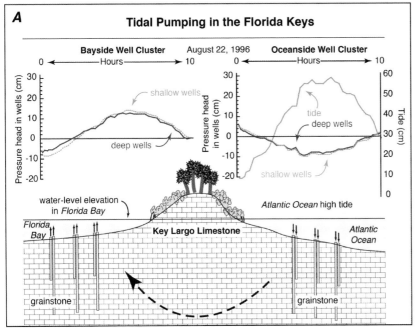

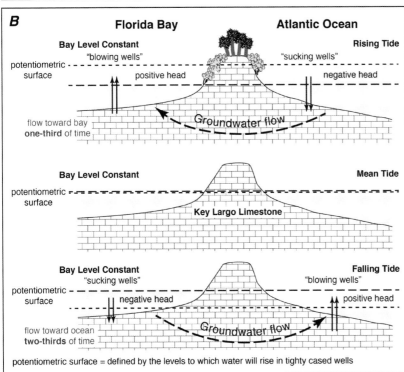

Figure 5.1. Conceptual models (not to scale) show bay-to-ocean groundwater flow in the upper Keys during rising, mean, and falling ocean tides. *A*, At high tide, pressure head is positive in bayside wells and negative in oceanside wells. This situation occurs about one-third of the time. Arrows above wells indicate direction water would flow if well caps were removed. *B*, Net flow direction is seaward about two-thirds of the time. Florida Bay has no tide per se. Bayside water level remains about 25 cm (10 in.) higher than average oceanside level. Bay "tidal" fluctuation was 3 cm (1.2 in.) over the 10-hour observation period. Pressure head at both well clusters was measured with an underwater manometer (Reich, 1996). Net groundwater flow throughout the Florida Keys, including in tidal channels and at the Dry Tortugas, is toward the Atlantic Ocean.

instigated coral demise in the Keys also produced simultaneous demise throughout the greater Caribbean Basin.

Pesticides in Paradise

Many citizens in the Keys suspect reef demise could be related to aerial and truck-based spraying of pesticides that began roughly in tandem with coral failure. Those citizens have long noted the effects of pesticides on butterflies. Ironically, without mosquito control, few people would live in the Keys, and the tourism economy would falter. Precious little research has been conducted to determine direct effects of mosquito control on reef corals. Interestingly, the National Park Service (Department of Interior) does not allow aerial spraying in the Everglades or on their properties within the Florida Keys. Both state and privately owned marine-research establishments that keep live marine organisms also remain exempt from pesticide spraying. The effect of pesticides in the Keys remains a much-debated topic. It is a double-edged sword.

On the other hand, similar decline has been occurring throughout the Caribbean, even in areas not subject to periodic pesticide spraying. The demise of coral reefs remains complicated by many competing hypotheses and proposed solutions, some of which are touched on here.

Sedimentation

Sediment produced and suspended during dredging, once prolific in the early days of Keys urbanization, is another cited cause of coral demise. Clearly, corals prefer areas of clear water, but documentation of suspended sediment causing death is skimpy, except in those instances where corals are completely buried. Water over the entire reef tract including the outer reefs remains extremely turbid for several weeks following hurricanes (Ball et al., 1967; Perkins and Enos, 1968). Other than physical damage, sporadic storm-derived sedimentation has had little discernible effect on coral health. Annual growth bands in 100-year-old head corals seldom show growth effects similar to those caused by low-temperature events (fig. 5.2A-B; Hudson, 1981). Most coral demise in the Keys occurred well after completion of the extensive dredge-and-fill projects of the 1950s and 1960s.

Coral Diseases

Coral diseases and so-called coral bleaching have recently become serious threats to corals in the Keys as well as throughout the Caribbean. Diseases became especially noticeable in the late 1970s and peaked in 1983, not just in Florida but also in most of

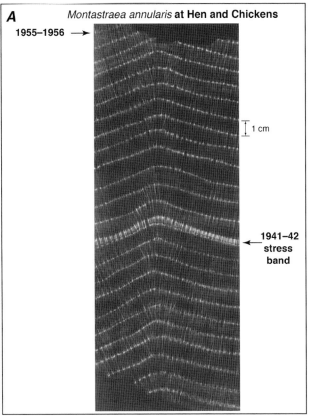

A *Montastraea annularis* **at Hen and Chickens**

1955–1956 →

1 cm

← 1941–42 stress band

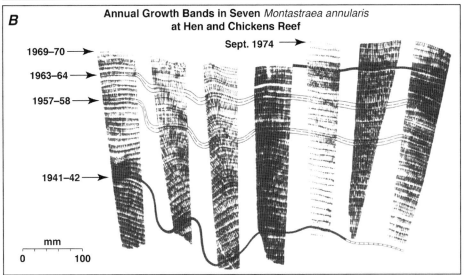

B **Annual Growth Bands in Seven** *Montastraea annularis* **at Hen and Chickens Reef**

Sept. 1974 →

1969–70 →
1963–64 →
1957–58 →
1941–42 →

mm
0   100

Figure 5.2. X-radiographs of coral core slabs show annual growth bands in *Montastraea annularis* from Hen and Chickens patch reef (from Hudson et al., 1976). *A*, Note abnormally wide stress band deposited by the coral in response to an unknown environmental perturbation during the winter of 1941–42. *B*, X-radiographs show correlation of banding in seven different *M. annularis* corals. Four at left are from heads killed by cold-water temperatures in the winter of 1969–70. Three at right are from corals that were alive and healthy when sampled in September 1974. The topmost stress bands joined by a dark line correlate with demise of the corals in the left X-radiographs. The stress band for 1941–42 indicates that conditions must have been as severe as in 1969–70 although these corals survived the earlier stress. The stress band for 1941–42 has been detected in other coral heads in the lower Keys. Other minor stress bands are very faint or do not correlate from coral to coral. Normal annual bands are correlative from coral to coral.

the Caribbean. In addition, the keystone herbivorous reef-dwelling black long-spined urchin *Diadema antillarum* that feeds on the algae that compete with coral for living space declined by about 90% throughout the Caribbean and Florida in 1983 (Lessios et al., 1984).

For the most part, Caribbean and Keys-wide coral bleaching began later, during the warm, hurricane-free, and unusually calm summer of 1987. Corals, especially head corals, turned white after expelling their microscopic symbiotic dinoflagellates called zooxanthellae. Scattered minor bleaching had been reported in previous years, but the major event began in 1987 and continues periodically to the present. Extreme bleaching events can result in coral mortality, but generally the symbionts, color, and health return with cooling waters.

Severe worldwide bleaching peaked during the El Niño Southern Oscillation (ENSO) event of 1997 and 1998. A sudden warming of surface waters in the western Pacific accompanied by oscillation of atmospheric pressure, sea-surface temperature, and wind causes these periodic events. The warming results in conditions that cause wind sheer over the Atlantic and that influence weather worldwide. The most recent El Niño to affect climate and coral growth occurred in 2015. Additional warming caused by these events is thought to weaken corals and accelerate their susceptibility to diseases. The influx of African dust also occurs mainly during El Niño events and likely caused additional threats to marine health in the past.

African Dust

The acroporid coral and *Diadema* demise that began in the late 1970s and peaked during the El Niño of 1983–84 was also attributed to the Caribbean-wide increase in airborne transport of soil dust from the Sahara and Sahel desert regions of North Africa (Shinn et al., 2000a). Joe Prospero at the University of Miami has monitored the flux of African dust across the Atlantic on the island of Barbados in the Lesser Antilles since 1965 and in Miami since 1974 (Prospero, 1999). His long-term data show that 1983 and 1984 were peak years of dust transport to the Caribbean and eastern United States (fig. 5.3). Initial study by USGS microbiologists and colleagues showed that the dust reaching the Virgin Islands (fig. 5.4A-B) contains abundant viable microbes and pathogens toxic to plants and humans (Kellogg et al., 2002; Griffin, 2007) and confirmed presence of pesticides (Garrison et al., 2014). Pesticides including DDT crossing the Atlantic have been known for many years (Risebrough et al., 1968).

The same dust brings phosphate and iron to the Amazon forest (Swap et al., 1992), and the iron in dust increases primary productivity in the open ocean (Young et al., 1991). Iron in dust has recently been implicated in the occurrence of red tides (Walsh and Steidinger, 2001). In addition, African dust contains arsenic and mercury

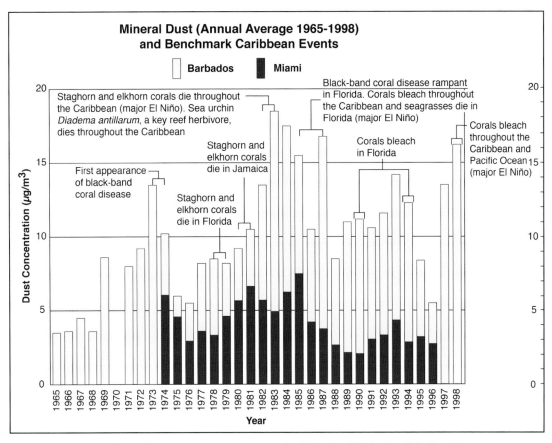

Figure 5.3. Average African dust concentrations measured in Miami and Barbados (Windward Islands) and major events that occurred in reef organisms in the Florida-Caribbean region from 1965 to 1998. Are they related? Data from 1965 to 1986 are from Prospero and Nees (1986). Data from 1987 to 1998 are courtesy of J.M. Prospero (director of CIMAS, Retired, of the University of Miami Rosenstiel School). African dust events can readily be observed on the Internet at NASA's website: http://earthobservatory.nasa.gov/NaturalHazards/ (accessed 11/17/16).

(Holmes and Miller, 2004). As scary as this might seem, the African dust hypothesis is not popular with many researchers and is generally not recognized by the public. Caribbean citizens in the main pathway, however, are well acquainted with African dust storms, and the breathing difficulties caused by dust are well known to Caribbean physicians, not to mention the mess dust leaves on boats and autos in its wake. Widespread ignorance of the dust hypothesis is thoroughly discussed in the memoir *Bootstrap Geologist,* published by University Press of Florida (Shinn, 2013).

Figure 5.4. *A*, Photo taken on 15 September 2000 shows period of normal visibility in the U.S. Virgin Islands. View is northward to Jost van Dyke, British Virgin Islands. Nearest islands are approximately 1 km (0.6 mi) distant. *B*, Photo of same view taken on 28 June 2000 shows obscured visibility when African dust enters the region. Courtesy of Virginia A. Garrison, USGS Research Ecologist (Ret.).

Sunscreens and Personal-Care Products

In recent years, laboratory studies have targeted a previously unsuspected coral foe. Recent research both here and abroad indicates that the UV filter oxybenzone and other chemicals used in sunscreens and sunscreen lotions and in personal-care products such as body fragrances, hair-styling goods, shampoos and conditioners, anti-aging creams, lip balms, mascaras, and insect repellants, as well as dishwasher soaps, dish soaps, hand soaps, and bath oils/salts can cause bleaching and deformation of coral larvae, or planulae. Such products and regular body oils enter popular coral reef areas either directly from swimmers and divers or via wastewater, transported in shower water, dishwasher water, and sewage. Because of their oiliness, these products tend to float, producing an oily sheen over reef areas. During spawning, coral planulae glide to the water surface to begin drift dispersal and maturation. At the surface, the planulae easily come in direct contact with the chemicals. Experimental contact with these chemicals, especially with oxybenzone (BP-3), demonstrates that extremely low concentrations (parts per trillion) can produce adverse effects on coral larvae. According to the US National Park Service, 6,000 to 8,000 *TONS* of sunscreen and personal-care products enter global coral reef areas *annually*. Expert scientists at the International Programme on the State of the Ocean think toxicity occurs at a concentration of 62 parts per trillion, equivalent to one drop of water in 6.5 Olympic-size swimming pools. Although banned in some reef parks in Mexico, at this writing, sunscreens containing oxybenzone remain legal in the Florida Keys.

Climate Change?

Currently, the most prevalent hypotheses for coral demise discussed in this chapter are related to climate change (warming seas), ocean acidification (alkalinity shift), and rising sea level. Notwithstanding the ill effects of coral bleaching in warming waters, corals along the east coast of Florida are thought to have responded to warming seas by expanding northward to higher latitudes (Precht and Aronson, 2004). The seemingly more insidious change may be caused by so-called ocean acidification. Marine organisms that precipitate calcium carbonate from seawater to build their skeletons or shells may be less able to do so now and in the future than in the past. In addition, less alkaline waters may corrode skeletal coral once the polyps die, regardless of the cause of death. That climate change could be a major threat to coral reefs is based mainly on computer projections and on laboratory tests using experimentally introduced high levels of CO_2 (carbon dioxide). These are just some of the main factors thought to explain recent coral death. Precht and Miller (2007) address more potential causes.

Results of future research will undoubtedly propose other currently unsuspected causes of coral death. The reader is reminded that whatever precipitated the recent demise in the Keys also occurred throughout the Caribbean at the same time. The reader is also reminded that regardless of reasons projected by various investigators, the detailed maps of bedrock topography and sediment/reef thickness indicate major changes in coral vitality also occurred periodically during the last 6,000 years of premodern coral history in the Florida Keys. Especially notice in figure 2.2 the areas where essentially no coral buildup occurred during that period of time. Could something in those areas be related to an unfavorable local microclimate? That noticeable changes have transpired recently, during the past few decades, is beyond doubt. But a question to consider is, are recent changes different from those that took place during the last 6,000 years? Additional research is needed to determine the source(s) of past coral die-offs. A potential method might involve ^{14}C-age determinations combined with calcium/strontium ratios of the dated material to determine water temperature around the time(s) of demise.

Photographic Documentation

Coral decline over the past three decades is well established both statistically and through observation by many researchers. The senior author has photographically documented demise of specific coral heads for the last 56 years. Photos can be seen at http://coastal.er.usgs.gov/african_dust/gallery.html (accessed 11/17/16). The time series of photos also demonstrates ecological shifts from coral-dominated to algae- and gorgonian-dominated habitat over that period of time.

The serial photographs (not all 56 years are shown) were taken at Grecian Rocks Reef (fig. 5.5A-B) and Carysfort Reef (figs. 3.11A, 3.12A). Both series began as part of a coral growth-rate study (Shinn, 1966). At Carysfort, an extensive reef with a major lighthouse, stainless-steel welding rods were driven into a large brain coral (*Diploria* sp.) in 1960 so that exactly 10 cm (4 in.) protruded above the living surface. The objective was to make regular measurements of the rod length as the coral grew up around it. Unfortunately, periodic measurements were not taken; however, by 1980 most of the pins were overgrown. When disease first appeared on the coral in the 1980s, bioerosion also began mainly as a result of parrotfish herbivory, and the pins reemerged. By July 2014, only two patches of live coral tissue remained, and one pin was left protruding about 15 cm (6 in.) above the dead surface. None of the large *Diploria* sp. is alive at this writing.

At the start of the experiment, a thriving thicket of staghorn coral (*Acropora cervicornis*) surrounded the *Diploria* head. After 1984, all the acroporid corals were dead. At this writing, there are a few colonies of staghorn that have been transplanted to the reef as part of an attempt to bring it back.

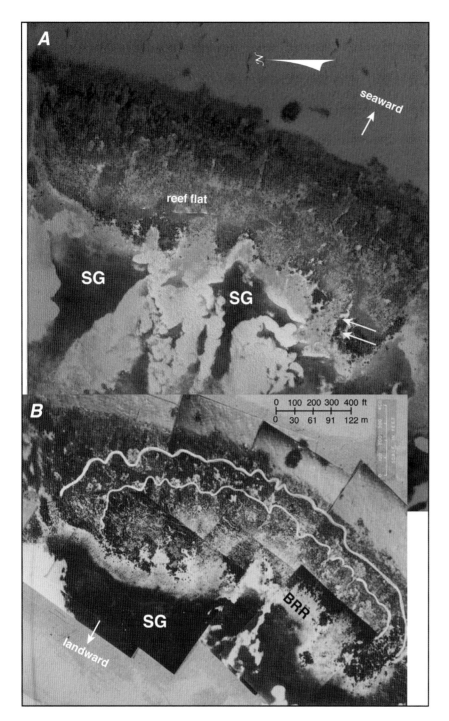

Figure 5.5. A, Google Earth satellite image (2013) shows Grecian Rocks reef. Yellow arrows indicate two sites of serial-photo sets available at website http://coastal.er.usgs.gov/african_dust/gallery.html (accessed 2/10/16). Note distribution of seagrasses (SG) surrounded by white reef sand in 3 to 4 m (~10 to 13 ft) of water landward of the reef. B, Uncorrected black-and-white photomosaic shows Grecian Rocks in 1959 (taken by Shinn from a small aircraft). Area between the two contour lines is the spur-and-groove zone from when *Acropora palmata* and other coral species were living. Area seaward of spur-and-groove zone consisted mainly of coral rubble. Area landward of spur-and-groove zone is reef flat, then composed mainly of in-situ live *A. palmata*. White area seaward of rubble zone is reef sand in 8 m (26 ft) of water. Note extensive seagrass (SG) west of backreef rubble (BRR). Most of the acroporid corals and many head corals were dead by 2015. Coral demise began in late 1970s and peaked in 1983. Acroporid demise also peaked throughout most of the Caribbean in 1983.

During the initial growth-rate study, published in Shinn (1966), most attention was focused on the Grecian Rocks[2] site where plastic bands had been attached to branches of *A. cervicornis* 10 cm (4 in.) from the growing tips. At the same time, living branches with bands from the same colony were transplanted to a site in Hawk Channel and a site in shallow water about 152 m (500 ft) from shore near Garden Cove on Key Largo. Simple maximum-minimum mercury thermometers in a plastic tube were installed at the Grecian Rocks and nearshore transplant sites. Coral-branch measurements were made with calipers between the plastic bands and growing tips each month for one year. Maximum and minimum temperatures that occurred since the previous measurement were also recorded. One significant result was that growth at the shallow, more turbid nearshore site equaled that of the offshore control site during the months when maximum and minimum temperatures were about the same; however, growth rate at that site had slowed, and the coral bleached during the summer when the temperature at the nearshore site rose to 30°C (86°F). The coral did not die, though, and color returned and growth resumed when temperatures cooled. However, when the temperature dropped below 13°C (55°F) in February, the branches died. Surprisingly, the annual growth rate of 10 cm/yr (4 in./yr) at the control site exceeded that previously determined at the Tortugas Carnegie Laboratory. A significant observation was that new branches appeared on the corals at the control site during February.

Why Florida reefs did not attain their full potential during the last 6,000 years of the Holocene remains unresolved. Growth rates are known for nearly all coral species, especially for the rapidly growing acroporids and massive head corals (fig. 5.6). For example *A. cervicornis,* often considered a "coral weed," grows at a rate of 10 cm (4 in.) per year at the control site and is known to grow much faster elsewhere in the Caribbean. On average, its branches produce about three new branches each year during February, and each new branch repeats the process the following year. Calculations indicate that a colony with only 10 branches has the potential to produce an unbelievable 56 km (35 mi) of branches in just 10 years! Stated another way, that rate would amount to more than 1 m (3 ft) of vertical growth in 10 years, far outstripping the past and present rate of sea-level rise (see fig. 1.5). Clearly, such potential growth is periodically curtailed by natural weather and disease events. Nonetheless, before the 1980s and the advent of recent coral diseases, countless *A. cervicornis* thickets at least 1 m (3 ft) tall were photographically documented in the Keys. In addition, *A. palmata,* the major constructor of reef spurs, grows at roughly the same rate as *A. cervicornis* and is much more wave-resistant. These rates are far faster than the known rate of Holocene sea-level rise, indicating such rapid rates of growth cannot be sustained. Storms break and scatter *A. cervicornis* thickets and branches hundreds of meters (Shinn et al., 2003). The more structurally competent *A. palmata* quickly reaches the sea surface to form reef flats, such as at Grecian Rocks (figs. 3.2A, 5.5A).

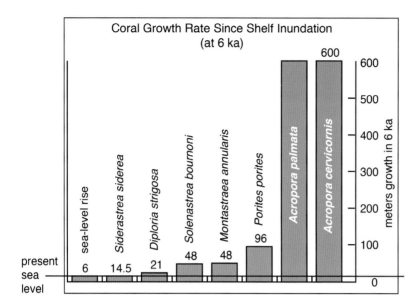

Figure 5.6. Histogram shows potential vertical growth of coral species over 6,000 years based on known annual growth rates. The shelf off the Keys has been flooded for ~6,000 years during the Holocene, yet none of the species other than acroporids has kept pace with rising sea level. Why they did not is not known.

Older reef flats, now dead and submerged, can be found on many reefs (fig. 3.10A-B). Such flats are excellent indicators of former positions of sea level but do not necessarily indicate prolonged stillstands, in the context of geologic time.

With the focus on geologic information and processes now provided, let's shift to the real world and offer tips for the reader who may wish to have a hands-on experience with lime mud and coral reefs on both sides of Key Largo. Let's take a field trip.

6

Key Limes, Hands On

See It in Person

As mentioned, geological and biological field trips have been conducted in the Florida Keys for many years. Such trips are somewhat of a rite of passage at universities with strong geology and biology departments. Many guidebooks have been prepared, beginning with those of Ginsburg when he led the modern carbonates Shell Research Group in Coral Gables in the 1950s. The Geological Society of America (GSA) published most of that field guide, his first, in 1964. In 1972, the Comparative Sedimentology Laboratory of the University of Miami, headed by Ginsburg, reprinted the GSA guide as *Sedimenta II*. Hoffmeister (1974) compiled an early comprehensive book for the nonspecialist on the geologic formations and geomorphic features of south Florida. Multer (1974, 1977) and Multer et al. (2002) prepared by far the most extensive geological guides, and Jaap (1984) penned an extensive biological guide to Florida coral reefs. Many major oil companies and universities have published their own geological guides (e.g., Harris, 1994; Harris and Moore, 1985; Eberli et al., 2014), but others likely remain proprietary. The authors of this book together with geologists Paul Enos, Robert Halley, and Randolph Steinen participated in and led numerous such field courses for scientists of the USGS, the American Association of Petroleum Geologists, and the Society for Sedimentary Geology (SEPM) (e.g., Shinn et al., 2000b). An extensive field guide summarizing then-15 years of research at the USGS Fisher Island Field Station was also prepared for the International Geological Congress in Washington, DC (Shinn et al., 1989). A more recent guide printed in color was prepared to celebrate the Centennial of the Carnegie Dry Tortugas Laboratory (Shinn and Jaap, 2011).

Unlike typical highway-based geologic guides with predetermined numbered stops, no two water-based field trips can be exactly alike. Water trips often require modifications on account of weather, boat availability, breakdowns, motel accommodations, and any number of other less important events such as box-lunch issues, sunburn, coral scratches, and *Diadema* punctures. Fortunately, no shark or barracuda attacks have occurred. Experience has been the best teacher, so here we briefly outline a typical field trip with some caveats.

A leading caveat is that if you as a leader wish your students and participants to see and appreciate lime mud, it is important that Florida Bay be visited first if at all possible. Florida Bay and its smelly mud will be a huge letdown to many if clear water, corals, and brilliantly hued reef fish have already dazzled participants. It's simple psychology. If you begin with swimming and snorkeling on the coral reefs, you may have great difficulty coaxing people out of the boat and onto a mud bank or a mosquito- and deer fly–infested mangrove island.

So with that said, and after making boat, box-lunch, and motel arrangements and obtaining dive masks, snorkels, fins, and inflatable safety vests, you are ready. Here are the classic stops for appreciating the smelly mud of Florida Bay. Don't forget insect repellant, tennis shoes, a long-sleeved shirt, a broad-brimmed hat, sunglasses, and sunscreen.[1] Sunburn can ruin a great educational experience. It could even make one miss the excitement of the coral reef.

Day One, Bay Side

Stop One: Make a stop on a typical mud bank in Florida Bay. Some years ago, the Park Service decided it no longer wanted students undertaking the classic stop, walking or crawling across Cross Bank (fig. 1.6A-B). One can see and experience a similar bank at an unnamed mud bank outside park boundaries just west of Key Largo (see location on fig. 1.6B). Start your trek on the north side of the bank and if possible take a short sediment core with a length of plastic tubing. It is also advisable to have a 3.6-m (12-ft) length of 1.27- or 1.91-cm (½ or ¾-in.) aluminum tubing to use as a probe. Use the probe to measure sediment thickness as you walk, crawl, squirm, slide, or swim across the bank. Be sure to observe the various sediment-producing organisms and the changes in sediment texture. Changes you will see were described in chapter 1 (figs. 1.9–1.12). Observe the different encrusting organisms attached to marine grasses. These organisms are among the many producers of the lime-mud sediment in which you are swimming or walking.

Stop Two: For your next stop, swim and snorkel in a so-called lake area in the bay (the local name for areas between mud banks, such as are shown in fig. 1.6B). Grab a handful of sediment from the bottom and test the bottom with the probe. Most likely the sediment will be less than 15 cm (6 in.) thick, and it will be shelly. Discuss why the sediment is different from that on the mud bank.

Stop Three: Walk on a mangrove island. The classic stop at Crane Key within Everglades National Park is out; however, you can see similar algal mats and mud cracks at Cotton Key (fig. 1.6A-B). The boat trip to Cotton Key will be on the Intracoastal Waterway and passes through Cow Pens Pass (fig. 1.6B) near the eastern end of Cross

Bank. Notice the dense mangroves on either side of the channel. In the 1950s and early 1960s, these mangroves were about 1 m (3 ft) tall. The pass is called Cow Pens because railroad workers once kept sea cows (as manatees were called back then) fenced in pens there to serve as food for the workers. With no refrigeration, turtles and manatees were kept alive until eaten. Old sour, a concoction made from Key limes (the small yellow kind) and salt, improved the taste.

Continue southward on the Intracoastal Waterway until you reach Cotton Key, which will be on your left (port side, in boating terms). You can go ashore on the northwest side of the island where there is usually an opening through the mangroves. Walk into the island to the narrow open area. Mangroves have been rapidly invading the open salty pond that runs the length of the island. Using a knife or small shovel, cut through the algal-matted surface to observe the laminations. You may find a storm layer that is stiff and about 2.54 cm thick (1 in.; fig. 1.9A-B). Hurricane Donna likely deposited it in 1960.

Stop Four: A pleasant way to end the day is to leave the bay through Snake Creek Channel (fig. 3.3A) and head out to Hen and Chickens patch reef on the landward edge of Hawk Channel. A small lighthouse is located on the seaward side of this patch reef. Alternatively, you can go to Cheeca Rocks (fig. 1.6A). Numerous mooring buoys are located on both patch reefs. At either place, you will see large coral heads, mainly species of *Montastraea*. More than half of these 100+-year-old heads were killed by cold water during the winter of 1969–70. This is an educational swim and a good way to end the day and remove any remaining sticky mud.

Day Two, Ocean Side

Stop One: Depending on which boat you engage, the suggested first stop should be either Rodriguez Key Bank or Tavernier Key Bank (see fig. 1.6A-B). Anchor off the seaward side of either bank, and swim or walk to the edge of the bank. Observe the band of finger coral *Porites divaricata* on the bank edge and then, higher up, the change from coral to coralline algae (species of *Goniolithon*, a red alga, fig. 6.1). Coral does not live on the shallowest part of this low ridge because the ridge is exposed during spring low tides; however, *Goniolithon* species can withstand exposure to the atmosphere. Depending on tide level, either walk or swim toward the island. As the water deepens by a few inches, notice the foot-high (30.5 cm) volcano-shaped sediment mounds. The mounds are excavations made by species of the ghost shrimp *Callianassa*. Adjacent to the mounds are 1- to 2-cm-diameter (0.39- to 0.79-in.) holes, usually several, that are intake/incurrent holes. The shrimp draws water into these holes with paddle-like swimmerets on the abdomen and then pumps it out along with sediment to produce the mounds. Insert your finger and notice the stiff slippery lining. The shrimp packs its mostly mud feces on the walls to keep it open. On top

Figure 6.1. One of the most common species of coralline red alga in the Florida Keys is *Goniolithon*.

1 cm

of the mound and falling along its sides are rods 12.7 mm long (½ in.) and about 2–3 mm (⅛ in.) in diameter. These are the muddy fecal pellets of *Callianassa*. The pellets resemble pieces of *Goniolithon*; however, they are soft and contain several internal longitudinal canals. Because of the internal structure of the pellets, their fossil counterparts were once thought to be a species of alga called *Favrina*. *Favrina* species are abundant constituents of Mesozoic (table i.2) limestone worldwide. Figure 6.2 shows a plastic cast made on Rodriguez Key Bank of a multilevel *Callianassa* burrow. Also notice the abundant seagrasses and the mermaid shaving brush *Penicillus*, scattered here and there (fig. 2.9). *Penicillus* species contain needle-shaped aragonite crystals that are released into the sediment upon death of the plant. The crystals are among the mud producers.

Stop Two: Continue by boat across Hawk Channel to White Bank (figs. i.1, 3.1). Stop and take a look in Hawk Channel if the water is clear. Anchor and swim on the white seagrass-free, rippled-sand area near the landward side of White Bank. The carbonate sand grains are well sorted and polished but are not true ooids. It is useful to swim over the landward side where the bank slopes down into the seagrass-covered eastern side of Hawk Channel. This is where the bank accreted landward during Hurricane Donna. You may find some hand-size fragments of this sand that have been cemented to form lime grainstone. The cement is aragonite.

Stop Three: A shallow well-marked coral patch within Hawk Channel and north of Stop Two is called the Cannon Patch Reef (fig. 3.1). At low tide, no more than

Figure 6.2. A polyester-plastic resin cast shows the deep burrow made by a species of the ghost shrimp *Callianassa*.

30 cm (1 ft) of water covers the coral heads. This is a good place to note the lack of branching acroporid corals. Also see if you can find the cannon reportedly located here.

Stop Four: Head seaward and across White Bank to the reef called Grecian Rocks[2] (figs. 3.2A, 5.5A-B). You can usually spot boats tied to moorings. Most mooring balls are on the landward side of the reef, which is about 762 m (2,500 ft) long. Approximately 12 m (40 ft) of Holocene sediment lie below the moorings. Cores also revealed a thin layer of peat below the sand and above the underlying limestone. Remember, this was swampy land about 6 ka.

Tie up to a mooring buoy and swim to the landward side of the reef flat. Waves may be breaking on the seaward side of the flat. You will see abundant 100+-year-old head corals, mainly species of *Montastraea* landward of the flat. Most likely you will not see many living acroporid corals but there are plenty of dead fragments littering the bottom. The horizontal surface of the reef flat formed by acroporids will be obvious. During high tide on a calm day, you can easily swim across the reef flat to the seaward side to observe incompletely formed spurs and grooves. Continue out to the seaward edge of the reef in about 7.6 m (25 ft) of water where coral rubble merges with carbonate sand. The sand continues seaward for about 1.6 km (1 mi) out to the shelf margin. The geologic cross section of this reef based on cores and ^{14}C dates was presented earlier (fig. 3.2A).

Stop Five: Now for the final stop you've been anticipating. Travel seaward and southward along the platform margin to either French Reef or Molasses Reef (fig. 6.3A-B). Molasses is easy to spot because of its lighthouse and many boats tethered to moorings. At either location, note the spectacular spurs and grooves. Try not to be distracted by the brilliantly colored reef fish!

Again, few if any living acroporids are present. Note the fire-coral coating that obscures the now-dead skeletal *Acropora palmata* that built the spurs. Dive down and observe *Halimeda* sand in the grooves. By now, you have had a long full day in the water. Time to return to shore, and enjoy dinner in a local restaurant with real Key Lime Pie for dessert!

Day Three, Onshore

Several land-based options are available for Day Three. Now that you have seen the Holocene coral patches and outer reefs, a visit to the quarry at Fossil Reef Geological State Park on Windley Key is indispensable (figs. i.1, 1.6A-B). It was here that limestone was quarried for construction of many early buildings in Miami. These include the main post office downtown and the courthouse in Coral Gables. Many homes were constructed from this rock, and many buildings in Florida are still being faced with a veneer of this beautiful coral-bearing limestone. Although active quarrying operations at Windley Key ceased in the mid-1960s, various cutouts to secure boats at waterfront homes still provide enough of this limestone to supply builders who use the stone for siding. Sawed slabs of this stone 5 cm (2 in.) thick are still available in Florida City near Homestead.

It is advisable to visit the small museum/learning center at Windley Key Park before walking into the quarry itself. The learning center contains many old photographs of quarry operations and a central glass-enclosed hands-on exhibit constructed by the authors and others showing a cross section extending from the quarry area out to the outermost offshore reef you saw the day before. The exhibit is built from the actual Substage-5e limestone quarried there, and the various fossil components are highlighted with descriptive text, laser beams, and underwater photographs. Patterned after a schematic cross section in a Florida Institute of Oceanography pamphlet, the exhibit includes the mud banks and islands of Florida Bay. Figure 6.4A-B, taken from the pamphlet, can be regarded as a summary figure illustrating the geology of the Florida reef tract. In addition, you can examine a core from a living Holocene *M. annularis* coral that began growing offshore at about the time the Pilgrims landed at Plymouth Rock. Harold Hudson drilled the coral and developed

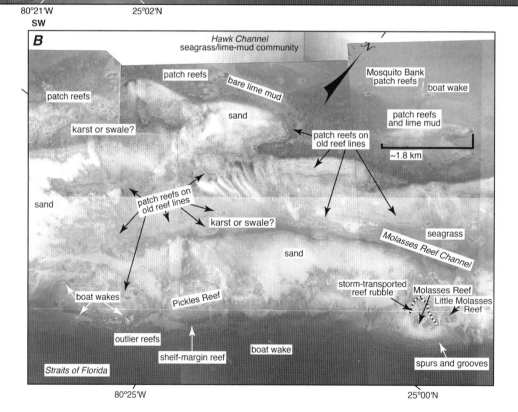

Figure 6.3. *A-B*, Contiguous aerial photomosaics (courtesy of Jim Pitts, 1975) show features and benthic environments on the outer shelf and margin seaward of Key Largo that can be observed on a self-guided field trip.

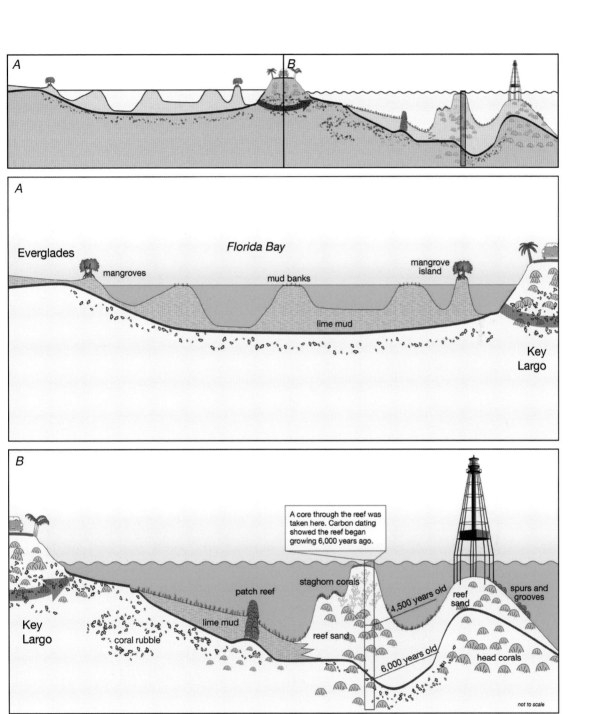

Figure 6.4. Highly simplified cross section from a Florida Institute of Oceanography pamphlet summarizes the geology of the Florida reef tract: A, bayside mud banks west of Key Largo and (B) tract east of Key Largo.

the attendant 3-m-tall (10-ft) core X-radiograph, on display at the museum, which shows the annual coral growth bands that resemble tree rings.

For a small fee, you can walk into the quarry and examine cross sections of the same species of corals and reef sands you have seen offshore (e.g., fig. 2.10). What you will not see is *Acropora palmata* or *A. cervicornis*, although some of the latter is present in this limestone elsewhere in the Keys. The best explanation for their absence equates to present conditions at places such as New Ground Shoal—no island barrier existed to protect the growing corals from cold Gulf of Mexico waters during Substage-5e time, when sea level was higher than the elevation you see in the quarry.

Most of the large corals exposed will have shallow, elongate, smooth, finger- and thumb-size holes scattered here and there. These are borings made by pholad clams, elongate bivalves that bore into limestone. Some holes will still contain the shells, and many will be partially filled with hard lime sediment. Notice the flat top of this internal sediment. These sediments are called geopetal fillings and, like a carpenter's bubble level, the flat horizontal upper surface indicates the up position. Such fillings are common throughout Earth's geologic history and are used to determine whether rocks have been tilted since deposition. The quarry is a good place to consider whether the Key Largo Limestone at this site was truly a coral reef or more like a sandbank with corals growing on it as they were being encased in accumulating lime sand, such as was discussed about White Bank.

Adjacent to the quarry is a nature trail through the hardwoods that once grew everywhere in the Keys. The trail begins next to the device on rails that was used to quarry the stone. On the trail, be careful of the poisonwood tree. Its black sap is very toxic to the skin. The trail will end in an adjacent older quarry where stone was first cut using wire cables coated with abrasives. Quarrying with abrasives is an ancient technique.

The Key Largo Limestone can be viewed at other locations in the upper Keys. Of particular interest is a complete cross-section exposure of the formation from ocean side to bay side in Adams Cut canal. The canal, located just north of the entrance to John Pennekamp Coral Reef State Park, is an artificial cut with vertical sides. Cross sections of large *Montastraea* heads that extend from water level to the land surface are exposed. Counts of growth bands in an X-radiograph of a large one (fig. 6.5), sampled from top to bottom, revealed the coral was 360 years old when it stopped growing toward the end of Substage-5e time. One can safely extrapolate that the 3.7-m-thick (12-ft, or 144-in.) section of Key Largo Limestone exposed in Adams Cut accumulated during a short period of only 350 to 400 years (Shinn et al., 1989). To put this rate of accretion into geological context, a sediment buildup of 2.54 cm (1 in.) per 1,000 years in the deep-sea environment is considered rapid. Unfortunately, homes and cutouts for small boats now obscure much of the Adams Cut site. Because

Figure 6.5. Cross section of a massive *Montastraea annularis* skeleton is exposed in man-made Adams Cut (also called Key Largo Waterway) through the Key Largo Limestone on Key Largo (upper Keys). The Key Largo Waterway connects Blackwater and Largo Sounds (fig. 1.1). The coral pictured grew on an unconformity (approximate location of white line at bottom; see note 3 for chapter 1). The reef top is 4 m (13 ft) above sea level at this site. To sample the specimen for X-ray analysis, four areas of the surface were coated in epoxy resin and cut with a power saw to remove V-shaped coral "cores" (red dotted lines; the V points into the coral). Annual growth bands in X-radiographs indicated the coral was 360 years old (Hudson, 1979).

of abundant construction, few locations remain from which to observe the geology here other than from a small boat.

A thick buildup of caliche, also now obscured by homes, occurs along the southeast side of the canal farther east where the land surface is about 0.6 m (2 ft) above water level. The laminated, brownish-red, very dense caliche coating was found to be ~5 ka at its base and around 0 ka at its smooth upper surface (Robbin and Stipp, 1979).

No trip to the Keys is complete without visiting the end of Highway US 1. Key West is prime tourist territory. You may enjoy the uniqueness of it all or find a dive-tour boat. Dive shops organize dive trips to offshore reefs including Sand Key Reef off Key West. The fast-catamaran day trip out to Fort Jefferson at Dry Tortugas is highly recommended. Along the way, you will pass near the Marquesas Keys and possibly over The Quicksands (figs. 4.1A-B, 4.3). Don't let the tour-boat leaders tell you the Marquesas were formed by a meteor strike!

Caliche/soilstone crusts abound on low-lying areas of the lower Keys oolite. A particularly good place to view the crust is on the left side of the road that leads to Seacamp on the south side of Big Pine Key. This relatively large open area is often flooded during high tides. One may even find shallow 10-cm-diameter (4-in.) core holes drilled by the senior author in the 1990s. That study revealed the presence of mercury within the laminated crust. The mercury came with atmospheric dust from North Africa and clearly predates modern habitation of the Keys. The rusty brown color is oxidized iron that is a common component of African dust.

Other Items of Interest—Historical Tidbits

Parks and Sanctuaries

Much of the area discussed in the upper Keys was originally designated as John Pennekamp Coral Reef State Park. It was named after John Pennekamp, a longtime editor of the *Miami Herald* newspaper. Pennekamp had been instrumental in acquiring the land for Everglades National Park in 1946. In 1976, a Supreme Court decision limited state waters on the east side of peninsular Florida to 5.5 km (3 nmi), leaving the most ecologically diverse coral reefs off the Florida Keys unprotected. Legal jurisdiction of Pennekamp Park was soon turned over to an emerging program in a new agency called NOAA (National Oceanographic and Atmospheric Administration, under the Department of Commerce). Enforcement was handled by the state and financed by NOAA. The new federally mandated park was called the Key Largo National Marine Sanctuary but is now part of the Florida Keys National Marine Sanctuary (FKNMS). Pennekamp Park still exists, but its jurisdiction also still extends only 5.5 km (3 nmi) offshore. An additional coral reef sanctuary was created in the lower Keys at Looe Key Reef. The Looe Key National Marine Sanctuary (see figs. 2.1, 2.2, 3.8) was the last bill signed into law by President Jimmy Carter. The Looe Key Sanctuary is also now part of the FKNMS.

On November 16, 1990, President George H.W. Bush signed a bill creating the 48,178-km (26,000-nmi) Florida Keys National Marine Sanctuary. The sanctuary size is presently 53,737 km (29,000 nmi) and extends from north Key Largo to the Dry Tortugas (fig. i.1). Jurisdiction includes all the waters and the seafloor out to a depth of 91 m (300 ft). Waters within 5.5 km (3 nmi) of the Keys, including some 4.8-km-diameter (3-mi) jurisdictional halos around small offshore sand islands in the lower Keys, are administered jointly by the State of Florida and the FKNMS.

In Air and on Ground

As you drive through the lower Keys, you will notice a silver blimp high in the sky over Cudjoe Key (fig. i.1). The tethered blimp, affectionately known locally as "Fat Albert," is officially aerostat radar. Its radar watches Cuba, looking for go-fast drug smugglers and small low-flying airplanes. Some say it beams old *I Love Lucy* TV programs to Cuba!

Be extra careful to observe speed limits, especially at night, while passing through the lower Keys. You will likely see dog-sized deer feeding alongside the road. Signs advertise the numbers recently killed by autos. Speeding-ticket fines contribute to the county economy. The deer, what might be called pigmy white tails, flourish by eating residential lawns and gardens, and the U.S. Fish and Wildlife Service protects their proliferating numbers. Not all residents are pleased with these cuddly deer, nor are they happy with growing flocks of feral chickens. The scruffy fowl frequent shopping-center parking lots on Big Pine Key (fig. i.1) and are considered a tourist attraction in downtown Key West. You literally don't need an alarm clock to wake at dawn!

Underwater Observation Chamber

In the early 1930s, a steel underwater observation chamber called the *Seaquarium* was placed in about 5 m (16 ft) of water off the southeast end of Grecian Rocks. The chamber consisted of a 2-m-diameter (6-ft) steel cylinder fitted with glass windows. The cylinder stood vertically and was entered from the top. The Labor Day Hurricane of 1935 knocked the chamber on its side. In the early 1950s, the Hempstead Brothers Salvage Company of Miami salvaged the chamber and used it as a mooring in Garden Cove on Key Largo. It was eventually taken to the scrap yard in Miami and sold as scrap iron. To our knowledge, there is no written record of its existence or the exact year it was installed. The senior author observed it underwater as a teenager and later when it served as a mooring. Learning more about this visionary device would be an excellent project for a modern-day historian.

Christ of the Abyss

In the mid-1960s, the Cressi family in Italy (of Cressi swim fins and other diving equipment) commissioned the casting of two replicas of a 2.7-m-tall (9-ft) 1,492.8-kg (4,000-lb) bronze Christ statue called *Christ of the Abyss* (*Il Cristo degli Abissi*) by Italian sculptor Guido Galletti. The original was placed in the San Fruttuoso Bay, Italy, in 1954. The first replica was placed near St. George's Harbor in Grenada in 1961. The second replica was donated to the Underwater Society of America to be placed somewhere in the Great Lakes. The Society decided that a more suitable place would be in the Florida Keys and offered it to the Florida Park Service. In 1965, when the area off Key Largo was a state park,[3] a group of volunteer divers from Miami installed the statue on a cement base in 8 m (26 ft) of water at what is now called Key Largo Dry Rocks. The ashes of one of those divers, Paul Damman, a longtime friend of the senior author, were recently spread at the site. The statue is featured on billboards and postcards, and literally hundreds of divers and snorkelers visit *Christ of the Abyss* each week.

Coral Castle

If you are homeward bound toward Miami, a visit to the Coral Castle might be of interest. Located just north of Homestead, the structure comprises numerous megalithic limestone blocks, each weighing several tons. Though commonly referred to as made of coral, the castle stones are actually oolite that may include localized concentrations of fossil shells and coral. See if you can find some of the cross-bedded layers mentioned earlier. The stones form the walls, carvings of celestial stars and planets, tables, rocking chairs, a bed, throne, an accurate sundial, a water well, fountain, bathtub, and a castle tower. The tower served as the actual living quarters of the eccentric builder. The stones are held together only by weight and skillful positioning, without mortar. A direct hit by Category 5 Hurricane Andrew in 1992 did not shift positions of any of the stones. The castle is noted for legends surrounding its construction that claim it was built single-handedly using "reverse magnetism" and "supernatural abilities" to carve and move the stones. An unconventional Latvian-American, Edward Leedskalnin (1887–1951), is said to have created the castle after his 16-year-old fiancée in Latvia jilted him just one day before the wedding. He made money by personally giving tours. The castle, listed on the National Register of Historic Places in 1984, remains open to tourists today.

Epilogue

The Future for Reefs and Research in the Florida Keys

Under modern climatic conditions, it is unlikely that eustatic sea level will fall either in our lifetime or in the next few thousand years. Considering its many fluctuations in the geologic past, however, the sea *will again fall* at some point in the geologic future as glaciers reform at Earth's poles. For geologists, sedimentologists, and coral reef ecologists, it may be useful to ponder the geologic effects of a stable sea level versus a rising sea level on environments of the Florida Keys and shelf.

A Stable Sea

Remember that modern reefs in the Florida Keys have evolved over a short period of time, in less than 7,000–6,000 years. During that time, sea level rose roughly 1.2 m (4 ft) per 1,000 years. Although most reefs have not kept pace with this rise in sea level, some have, those that are named, that are well marked, and that have grown upward at least 12 m (40 ft).

If sea level were to remain stable for, say, 1 million years, a geologically short period of time, and oceanic currents were to remain as they are today, the entire region from the southern boundary of the Everglades to reefs at the edge of the present shelf margin would fill with sediment and coral debris. This would be particularly true off the upper and northern middle Keys, where prevailing wave energy is onshore. The area would probably become a huge tidal flat or swamp. The filling of the shallow Florida Bay and Hawk Channel lagoons would block the source of potentially lethal cold water from the Gulf of Mexico. Corals would recruit to exposed Pleistocene and Holocene hardbottom areas on the shelf margin and on the outlier reefs and would grow unimpeded, bathed only by clear, warm waters of the Gulf Stream. With no backreef accommodation[1] available, backstepping would cease, and reefs would accrete upward and begin stepping seaward into the Straits of Florida. Depending on climatic conditions, the new land might be covered by migrating sand dunes.

Because of southwestward current flow off the lower Keys, sediments would accumulate on the southwestern shelf area more slowly.

Rising Sea Level

If, on the other hand, sea level were to continue rising, as is documented to be happening today, then different changes would occur on the shelf. (Remember, the oolite and coral reefs that built the Florida Keys are indisputable proof that relative sea level was 6 m [20+ ft] higher 125 ka than today.[2]) The existing Keys would become the nuclei for new coral growth as they have in the past, and the coral species that become established will likely be the same massive head-coral species that grew there 125 ka. The new growth would be separated from the old by a brown caliche crust similar to those separating earlier Pleistocene limestone units beneath the Keys. Coring and radiometric dating of corals, peat, and caliche have shown that south Florida and the Bahamas have undergone at least seven such cycles of sea-level rise and fall in the not-so-distant past. A caliche crust caps five of the seven marine-rock units formed by each fluctuation cycle and divides the youngest Pleistocene limestone (77.8 ka, a coral age from below Carysfort Reef) from the oldest Holocene accretion (9.6 ka, a coral age from a core drilled on Tortugas Bank west of the Dry Tortugas[3]). Surely, processes of the past and present are keys to those of the future.

Across the planet, corals have survived for the last 500 million years of Earth's history. In so doing, they have endured the oscillating vagaries of an impatient sea through time, and they will continue to do so well into the future.

The Future of Reef Research

It goes without saying that the improvement in diving made with scuba in the late 1940s advanced many forms of shallow-water marine research, especially coral reef research and underwater photography. Scuba replaced cumbersome diving helmets like those used in the early 1900s at the Tortugas Carnegie Laboratory. Unquestionably the big push for scuba diving and underwater research can be attributed to Jacques-Yves Cousteau's early usage in his 1953 film *The Silent World* and later in his widely popular TV series. What remained for the coral reef geologist was to find a way to look *inside* a reef as well as *below* the seabed. In so doing, the researcher would unlock many mysteries of the reef.

Figures 6.4A and B pictorially summarize the most recent geologic history of the south Florida shelf. Relative to geology of the Florida Keys and coral reefs, this book addresses answers, as currently understood, to many questions. What made the Key Largo Limestone? When did it live? How thick is it? What is it sitting on? Why did coral grow where it did? How fast did it grow? How and why was antecedent topography a factor? What made the Miami Limestone? Why did tidal bars form at each end of the reef? What kinds of grains formed the tidal bars? How did the grains become rock? What is the purpose of spur-and-groove systems? What makes them resemble a toothed comb? What are the kinds of reefs along the Florida reef tract? Why was position of sea level important? What types of evidence indicate that positions of sea level in the recent past were lower than at present? What evidence indicates sea level was higher than at present at least once in the recent past? How were natural tidal passes between the Florida Keys an influence on coral location? How did the creation of Florida Bay affect reef growth? Why is the fine-grained mud in the bay important? Why are algal mats and mud cracks on bay islands important? How do some mud banks keep above a rising sea? What makes Florida Bay water harmful to offshore reefs? Why is the direction of groundwater flow predominantly toward the Atlantic Ocean and the reefs? How has the human populace in the Keys impacted the coral reefs? How are winter storms and summer hurricanes important to vitality of coral reefs and to sedimentary processes? What is caliche? When, where, and how does it form? What gives it color in the Florida-Caribbean region? Why is it important in Florida? What organisms produce marine sediments? What is the primary sediment producer in The Quicksands? What kind of geologic/morphologic feature results from reversing tidal currents? What is beachrock and how does it form? What are birdseye structures and how do they form? What do they indicate about the environment when they formed? What are planktic Foraminifera? How do they tell us about past ocean temperatures and positions of sea level? What are benthic Foraminifera? Why are they and other microscopic calcium-carbonate organisms important to reefal limestone formation? What do caliche, sunken beachrock, or mangrove peats in an offshore core indicate? Why is African dust important to the marine ecosystem? Yet other mysteries persist, one of the biggest being the question of possible asymmetric peninsular subsidence and its timing on the geologic timescale.

When the authors began work in the Florida Keys in the early 1960s, the newest most technologically advanced research tools for subsurface work (below the seafloor) were mini-sparkers followed by boomers. They were analog seismic devices that recorded data on paper charts. In the 1980s, Loran-C[4] became available for navigation. The global positioning system (GPS) came along later to improve mapping and location accuracy. Aerial photography acquired in shallow water on cloudless days was crucial for interpreting bottom features, geomorphologies, and habitats. Satellite imagery greatly enhanced our knowledge of Earth systems and processes. Data were eventually recorded digitally.

A major advancement in reconstructing the geologic history of coral reefs was development of diver-operated hydraulic coring devices. In the early 1970s, Ian Macintyre and Walter Adey constructed the first such device. Soon after, the senior author and Harold Hudson at the USGS built a similar device (see fig. 2.3), thus launching three decades of coral reef coring for the USGS in Florida, the Bahamas, Belize, Puerto Rico, Enewetak Atoll in the Pacific Ocean, and ancient algal reefs up-lifted in the Sacramento Mountains of New Mexico. Improvements in the core drill have been made over the years, and today numerous water-monitoring wells have been installed both on- and offshore from the Florida Keys. Coring remains a time-consuming and laborious task but is critical to obtaining the geologic record stored in the recovered sediment and limestone samples.

Water temperature was initially measured with mercury thermometers, and for longer records, maximum/minimum thermometers designed for gardeners were sealed in watertight plastic tubes. They did not record continuously. For continuous records, the next advancement was a rather expensive commercial apparatus consisting of an 8-mm time-lapse surveillance camera focused on a dial-type oven thermometer and a cheap wristwatch. The entire package including batteries was sealed in an unwieldy watertight container. These cumbersome devices were hard to hide from curious divers, and battery life constrained recording length. Reading the developed movie film under a microscope was time consuming. Data were recorded with typewriters. Fortunately, researchers now have access to tiny inexpensive digital devices that can be placed in monitoring wells or hidden in coral crevices. These instruments record for many months, and data can be downloaded directly onto computers. Needless to say, bulky underwater-film cameras also evolved quickly into digital equipment.

The future for research on coral reef and sedimentary processes should be much brighter than in the early days. Remote sensing, such as the use of aerial photography, has evolved into a number of different technologies and functions applied both in situ and from afar. Lidar (light detection and ranging) operated from aircraft uses high-intensity laser light beams that easily penetrate relatively clear water to depths in excess of 18 m (60 ft) or more. The signals that bounce back measure the land or seafloor elevation. When received, the elevation data are used to construct GPS-controlled topographic and bathymetry maps of the land or seafloor. Virtually all reefs off the east coast of Florida have been or are now being mapped with lidar. Such features as dredged areas and sewage-outfall pipes and even spur-and-groove systems are clearly visible. Lidar in tandem with other new techniques is the wave of the future. For example, multibeam echo sounding, sidescan sonar, swath bathymetry, and the use of remotely operated vehicles (ROVs) and autonomous underwater vehicles (AUVs) are revolutionizing how we view and map details of the seafloor, regardless of depth. These techniques extend far beyond coral reef research and include the real-time and real-world searches for shipwrecks and downed aircraft at depths beyond those of human endurance.

With the new equipment, even individual head corals can be viewed with aircraft-operated lidar. Google Earth digital satellite imagery has improved so much that specific head corals and other bathymetric features can now be readily observed in imagery captured from space.

Another remote-sensing system called ATRIS (along-tract reef-imaging system) is also currently in use. It consists of a towed downward-looking video camera that images the in-situ seascape and benthic habitats down to 27 m (88.5 ft) below the sea surface. The system allows continuous and simultaneous acquisition of geolocated, color, digital images and corresponding imaging distances. The imagery nevertheless requires online visual examination and interpretation to define types of benthic habitats and local biogeologic processes (Lidz et al., 2008b). Recently deployed at the Dry Tortugas, ATRIS technology has documented renewed presence of healthy *Acropora cervicornis* colonies and adult *Diadema antillarum* that may represent a fledgling recovery in acroporid growth and herbivorous-urchin repopulation (Lidz and Zawada, 2013).

It remains to be seen what new devices are just around the corner for future carbonate sedimentologists and coral reef-and-limestone studies. Perhaps even newer instrumentation will be deployed from space. Unfortunately, the downside is that all this technology tends to take researchers out of the field and water and into air-conditioned offices far removed from the real world. Our fear is that reef scientists will gradually lose the hands-on approach needed for appreciating and accurately resolving the natural-history mysteries that still remain. Increasingly, various well-meaning groups are pushing for more investigative regulations, and bureaucracies charged with creating and enforcing new regulations may lack the hands-on experience needed for well-informed and well-balanced decisions. When the authors first began scientific investigations in the Florida Keys in the 1960s, no permit to conduct research was required. That changed with creation of sanctuaries, preserves, refuges, and parks. Tighter rules and regulations and elaborate permitting systems eventually evolved. Necessity for research permits continues to escalate, and permits for some activities may no longer be obtainable at all, especially if the work involves an endangered species or its designated critical habitat. Even simple requests, such as the nondestructive act of collecting small quantities of bottom sediment for chemical or constituent analyses, require completion of long multipage online forms. Unfortunately, such well-intentioned procedures can slow the pace of research, or worse—by causing reef scientists to conduct their studies, at greater expense, outside the Keys, in foreign waters so to speak. Results of those studies cannot be extrapolated to south Florida reefs because conclusions drawn from elsewhere might not correctly characterize conditions off the Keys. The risk also exists that such procedures will encourage researchers, especially students, to restrict their investigations to highly statistical

Figure e.1. Sunset enhanced by an African dust event in June 2015 as seen from west side of Lower Matecumbe Key. Courtesy of Martin Moe, fishery biologist (Ret.), and marine-fish aquaculturist.

remote sensing and modeling studies carried out on computers and in laboratories. In our experience, the needed answers to natural-history unknowns are to be pursued and discovered in the field. As geologist Francis Pettijohn emphasized 60 years ago in his SEPM Presidential Address, "In Defense of Outdoor Geology," the real problems, and their solutions, still lie in the field. And if one never ventures outdoors, just look what might well be missed (fig. e.1)—a stunning natural event, compliments of rays of the setting sun reflecting off atmospheric African dust!

NOTES

Introduction

1. The William H. Twenhofel Medal is the highest award bestowed by the Society for Sedimentary Geology (SEPM) in recognition of "Outstanding Contributions to Sedimentary Geology." By 1985, when R.N. Ginsburg received the Twenhofel Medal, he had already influenced the course of carbonate geology. Bob had the rare ability to encourage those around him to think, to ask the "So what" question and, "What does it mean?"

Attending the University of Chicago on a G.I. Bill, Bob could go to the cafeteria and rub elbows with Enrico Fermi, Harold Urey, or Bill Libby, the true giants in their field. In addition, there were a plethora of great geologists and teachers around him, including Francis Pettijohn, Heinz Lowenstam, Cesare Emiliani, Julian Goldsmith, J Harlen Bretz, and Marvin Weller. These professors and their many students soon became known in the field of geology as the "Chicago Mafia." When Bob finished his PhD in 1954, he joined Shell Research and surrounded himself with those who would later become known as the "Shell Mafia." Ginsburg pioneered the concept of teaching field trips in south Florida and the Florida Keys. Many hundreds of geologists from all over the world have participated to learn sedimentary processes from the ground up. Most have used that knowledge in their fields of research, and they in turn have led their students on learning trips to the Florida Keys. The authors were two of the more recent participants and would later become true collaborators with Bob in definitive studies. We dedicate this book to the memory of our colleague and friend, Robert N. Ginsburg (1925–2017), and to the great knowledge he so inspiringly bestowed on us all.

2. Figure i.1 shows the location of NOAA's newly named Captain Roy's Reef, a patch of large coral heads of *Montastraea annularis* marked with a submerged memorial monument off north Key Largo. Captain Roy and his two charter vessels, the *Sea Angel* and *Captain's Lady*, were instrumental in scores of cruises during the USGS 1974 to 1997 acquisition of scientific data discussed in this book. The figure also shows the generally margin-parallel nature of Pleistocene and Holocene geomorphic trends seaward of the Keys.

3. Commonly known as the West Indian top shell, *Cittarium pica* is of interest because the gastropods live in the rocky intertidal zone throughout the Bahamas. Nowhere do they live, nor have they lived, in the Florida Keys today or during the past 10,000 years. In fact, the last time they lived in Florida was about 125,000 years ago during the Pleistocene Epoch, when the sea was several meters above present level. Their presence in earthen Tequesta mounds in Florida indicates that those Native Americans were a seafaring lot.

4. It is claimed that Christopher Columbus first encountered the New World at the island of San Salvador in the eastern Bahamas in October 1492. Although he made four voyages to the west and discovered many Caribbean islands, he did not find the Florida Keys. They were discovered 21 years later in 1513 by Spanish explorer Juan Ponce de Leon, who named the island chain Los Martires or the martyrs because, from offshore, the reefs and islands resembled suffering men. He likely saw the Keys after a devastating hurricane that had stripped the trees bare of leaves, or maybe he saw the upraised limbs of elkhorn coral exposed above the water line. The mainland to the north apparently looked more luxurious than the Keys, and he called the peninsula "La Florida" (Land of Flowers).

5. Wreckers were marine salvagers who rescued people and cargos from ships stranded on coral reefs. Wrecking first began in the Keys in the 1500s with Native Americans plundering cargos, but it was not until the 1800s that the industry became a more-or-less honorable profession. Law and order came in 1829 with establishment of the wrecker's court in Key West. Judgments of the U.S. District Courts followed Admiralty law. Cargo salvaged from wrecked ships was brought to Key West, where it was appraised and auctioned. Payments to captains and crews and shipowners required complicated judgments based on various percentages for all parties involved. The U.S. Government also got its share as duty on imported goods. Judgments and money distributed were not always on the up-and-up. Cargo salvaged included rich French silk fabrics, Spanish shawls, European clothing, jewels, sewing machines, lumber, coal, scrap iron, bales of cotton, and casks of wine. Modern-day booty includes bales of pot.

6. More than 1,000 shipwrecks dating to the 1500s, 1600s, and 1700s lie off the Florida Keys. Most are Spanish, but some are Dutch, British, and American. They wrecked as a result of storms or illness that disabled the crew.

7. More-civilized industries included cigar making, pineapple farming, and shrimping. Also noteworthy is that Key West has been home to famous writers, the most familiar being Tennessee Williams and Ernest Hemingway. Today, the Hemingway house, replete with its many well-cared-for multi-toed cats, is a historic residence, and Ernest Hemingway lookalike contests are annual events on the island. The Key West Naval Base often hosted visits from President Harry Truman. Truman Avenue and the Truman Annex remain parts of the base today.

8. Henry Morrison Flagler, an American tycoon and partner in Standard Oil with John D. Rockefeller, extended his railroad from New York to Palm Beach on the east coast of Florida to transport tourists and citrus, but a devastating winter freeze in 1894–95 wiped out most of the orange groves in central Florida. Oranges farther south survived. That and a fabled publicity stunt motivated him to extend his tracks to Miami, then a small settlement of fewer than 50 inhabitants! The often-repeated story is that Julia Tuttle, who ran a trading post in Miami, sent Flagler a box of orange blossoms to demonstrate that oranges farther south had survived.

At the time, coal was needed in Key West to refuel steamships headed to or from Cuba, New Orleans, and New York. This need provided motivation for Flagler to extend the line, known as the Florida East Coast Railway, to Key West. That extension would later be known as the Florida Overseas Railroad. In spite of delays and three hurricanes during construction, the line finally reached the southernmost city in 1912, eventually to be destroyed by 322-km/hr (200-mph) winds and a 6-m (20-ft) surge from the great Category 5 Labor Day Hurricane of 1935. The railroad was never reconstructed. A limestone memorial to the more than 400 people who lost their lives stands at Mile Marker 82 on Highway US 1. Many of the numerous bridges and adjacent mud-filled areas built to support the railroad tracks survived and later became the foundation for a widened US 1.

9. Key West prospered in the 1800s, but that prosperity would soon decline with the placement of light ships and lighthouses and the advent of the American Civil War. Lighthouses first

placed on shore were established at Cape Florida on Key Biscayne off Miami, at Key West, at Sand Key off Key West, and at Garden Key in the Dry Tortugas in the 1820s. The building of lighthouses on coral reefs did not begin until 1852.

Chapter 1. About the Keys: Processes I

1. See the Abbreviations section for English- and metric-system abbreviations for distance, size, and a past age or point in geologic time.

2. We use the classification of Robert J. Dunham (1962), modified to include only the most frequently used terms. Several classifications exist, but his is the least complex and most intuitively simple. Petroleum geologists commonly use it for describing cores and rock cuttings as they come to the surface during well drilling. Although intended for rock, the classification can also be used as if the sediment were rock. It is based on grain size at the time of deposition.

- Mudstone contains less than 10% grains (usually assessed by area in cut or thin section; see note 4 for chapter 2), supported by a lime mud.
- Wackestone consists of more than 10% grains, supported by a lime mud.
- Grainstone lacks mud and is grain supported.
- Packstone contains lime mud and is grain supported.
- Boundstone describes sediment in which the original components have been organically bound together during deposition.
- Framestone is a solid calcareous or siliceous framework that is maintained by an organism such as a coral or sponge.
- Crystalline carbonate does not have recognizable depositional structures.

3. An unconformity occurs when there is a gap in time (hiatus) in the rock record represented by missing strata or by a rock unit that is overlain by another that is not next in stratigraphic succession (see note 1 for chapter 2).

4. Birdseye structures, known as fenestrae in European geological literature, occur in two divergent forms: (1) as elongate horizontal voids, or (2) as spherical bubble-shaped voids. Both forms can also occur together in the same layer of sedimentary limestone. Horizontal shapes are formed in lime mud parallel to bedding planes through repeated shrinkage and expansion during desiccation. They vary in length up to 1 cm (0.39 in.) or more. Bubble shapes are usually about 1 mm (0.039 in.) in diameter, and studies in the Florida Keys showed they in fact originate as gas bubbles in soupy lime mud. Birdseye structures occur throughout the geologic column (i.e., the geologic record), mainly in light-colored fine-grained dolomite or lime mudstone. The structures are usually filled with calcite or anhydrite. Both forms are generally associated with mud cracks, vertical desiccation cracks, algal laminations, and organosedimentary structures produced by metabolic activity of microorganisms. Together, these characteristics are considered reliable indicators of the intertidal to supratidal zone and thus of sea-level position. (The supratidal zone is just above high-tide level.) Birdseye structures were first named because of their similarity with features found in wood slabs of Birdseye Maple.

5. Because of advancements in the field of genetics, new DNA evidence continually leads to the reclassification and renaming of plants and animals. Here we refer to a coral, now placed in the genus *Orbicella*, that built a large portion of Pleistocene and Holocene reefs in the Florida Keys. We call it by its former genus name, *Montastraea*, assigned before DNA analyses were possible, because this name is used throughout the geologic literature to date. Explanation courtesy of Ilsa Kuffner.

Chapter 2. Results of Data Gathering and Mapmaking: Processes II

1. The term stratigraphic refers to strata having been deposited in geologic order, with the oldest layer, or stratum, being on the bottom and the youngest being on top. For a complete stratigraphic or geologic record, strata representing continuous geologic time must be present with no hiatal gaps of missing rock. In the Florida Keys, the stratigraphic record of the past 125,000 years is in geologic order but is incomplete, in that the strata accumulated during specific time periods and time gaps are present in the rock record.

2. The Holocene isopach map, specific cross sections, and specific types of accretions show sites and types of localized processes and effects of fluctuations in sea level and water temperature on carbonate environments and corals. The following (examples in parentheses) are discussed in the book, either before, or in, chapter 2 or later chapters.

(a) Sediment accumulates in bedrock depressions (north of New Ground Shoal, and in the sinkhole off Key Largo; fig. 3.11A).

(b) Sediment accumulates behind rock barriers (east of Halfmoon Shoal in the Gulf of Mexico).

(c) Sediment does not accumulate on elevated bedrock highs, especially those washed by strong currents (seaward face of the shelf-margin reef).

(d) Sediment does not accumulate in bedrock depressions flushed by strong reversing tidal currents (Boca Grande Channel).

(e) Sediment in the lower Keys is transported offshelf (buried deep reef at Looe Key Reef; >40-m-thick [>130-ft] sediment wedge southwest of the Marquesas Keys; relatively empty backreef troughs behind the lower Keys outlier reefs; thick sediment accrual off west end of the Marquesas-Quicksands ridge).

(f) Sediment in the northeast is transported landward (sediment-filled backreef trough behind The Elbow and Carysfort Reef; buried incipient outlier reefs on terrace seaward of The Elbow).

(g) Keys-wide, sediment and reef debris are transported landward from major shelf-edge reefs by storms (Grecian Rocks, Sombrero Key Reef, Looe Key Reef, Sand Key Reef).

(h) Large sand bodies can form in place (White Bank off Key Largo, and in The Quicksands on the Marquesas-Quicksands ridge, which is isolated by deep water and strong currents).

(i) Holocene sands are accumulating on top of cemented Pleistocene sands in backreef topographic lows (ridge-and-swale geometry near Marker G).

(j) Holocene corals are recruiting to elevated Pleistocene coral ridges (ridge-and-swale geometry near Marker G, and at Grecian Rocks).

(k) Mangrove peat submerged beneath corals or coral debris indicates ancient pre-reef shorelines (Davis, Alligator, and Looe Key reefs).

(l) Laminated-calcrete caps on submerged or buried marine sequences indicate times of subaerial exposure and sites of once-dry land (on Pleistocene accretions at depth, and separating the last Pleistocene buildup from those of the Holocene).

(m) High-density annual bands in coral cores, when counted down core from year of recovery, indicate precise years of cold-water stress (Hen and Chickens Reef).

(n) Blackened limestone pebbles at buried subaerial-exposure horizons and in modern limestones indicate periods of wildfires kindled by lightning strikes (quarry tailings at Big Pine Key).

3. The pie chart on the habitat map shows areas and percentages of total area mapped for habitats and urban features (roads/airports). Acreage of each map unit was derived from the original benthic-habitat ArcInfo coverage. The coverage was converted into an ArcView shapefile and re-projected from geographic (latitude/longitude) into UTM coordinates. The re-projection converted acreage from a measurement of degrees into area (m^2 and km^2). The Crystal Report wizard in ArcMap was used to generate the final map-unit sums in m^2 and km^2. The sums were entered into an Excel spreadsheet to calculate percentages and produce the pie chart.

4. A thin section is a piece of rock or mineral, or a carbonate shell, skeleton, or limestone fragment that is mechanically ground to a thickness of approximately 0.03 mm and mounted on glass as a microscope slide. This process renders most such material transparent or translucent, thus making it possible to study its optical properties. Grains of various carbonate materials have unique and distinct optical properties that permit delineation of grain origin when viewed in thin section under the cross-polarized light of a petrographic microscope. Random point-count transects are made across a slide from top to bottom to account for any sorting by size during preparation. Each slide is advanced by three clicks until the lens crosshairs land on a grain, which is identified if possible, and counted. A grain, including any large ones, is counted only once. The thin-sectioned carbonate grains are identified based on previous experience and according to the carbonate petrography manual compiled by Scholle (1978).

A petrographic or polarizing microscope is one that uses polarized light and a revolving stage for analysis of petrographic thin sections. Two nicols (a nicol prism or Polaroid), one above and the other below the stage, polarize and analyze the light. The stage rotates about the line-of-sight axis. Polarized light is light that has been changed by passage through a prism or other polarizer.

5. *Halimeda* grains dominate The Quicksands (Hudson, 1985). Sedimentary data for The Quicksands were not included in the map of sediment compositions because the data are not contiguous with the Keys data. Sediment sampling and analyses have not been conducted between Key West and the Marquesas Keys. Percentages of *Halimeda* and coral grains in The Quicksands are contoured in Shinn et al. (1990).

6. Bioerosion is the biological breakdown of hard mineral materials (e.g., coral skeletons, bivalve or mollusk shells, or hard sedimentary rocks) by organisms, such as boring clams and sponges, and by the rasping and grazing of algae by urchins and some fish species, e.g., parrotfish.

Chapter 3. Major Geomorphic Topographies

1. During Hurricane Donna, lime mud settled in high-energy areas normally occupied by coarse-grained sediment. This mud was resuspended in the water column by waves each time wind velocity exceeded 16–24 km/hr (10–15 mph). Sediment distribution eventually returned to pre-storm locations, and much mud settled into crevices and voids within the reef framework. Mud was also transported into deeper water off the shelf. Abundant lime mud flowing off the shelf was initially transported northward with the Florida Current, creating layers several centimeters (inches) thick in coral areas farther north off Fort Lauderdale. The same process was observed following hurricanes in 2005.

2. As of July 2015, no living *A. palmata* could be found at either Grecian Rocks or Carysfort Reef.

3. Global positioning system (GPS) technology has rendered lighthouses virtually obsolete.

4. Marine transgressions occur during interglacial periods when sea level is rising or sea stands are high and coastal lands become flooded. Marine regressions occur during glacial periods such as the Pleistocene Ice Ages, times when sea level is falling or stands are low and coastal lands become exposed to the atmosphere. Coral reefs can accrete on marine regressions

if conditions are suitable and the rate of sea-level fall is sufficiently slow. For example, comparison of the depth of the oldest Pleistocene coral dated (77.8 ka) in Florida relative to the $\delta^{18}O$ sea-level curve (fig. 1.1) correlates with the beginning of falling sea level at the end of Substage-5a time.

Chapter 4. Western Terminus of the Reef Tract

1. Keystone vugs are formed when trapped air bubbles typical of the beach swash zone become preserved at the same time the sediment becomes cemented to form beachrock. Sand grains at the top of trapped bubbles form a "keystone arch" that allows the voids to persist long enough to be preserved when the surrounding grains become cemented together. Such features are excellent indicators of intertidal conditions.

2. As an example, the 122-m-long (400-ft) steel ship *Valbanera* transporting 488 Spanish cigar workers from Spain to Havana grounded on Halfmoon Shoal during a hurricane in September 1919. The ship could not enter Havana Harbor and was headed for Key West. All hands were lost. It was one of the worst disasters in maritime history at the time. Shown on marine charts and marked by a wreck buoy, the ship lies half buried in The Quicksands at Halfmoon Shoal.

3. The north edge of the Marquesas-Quicksands ridge is within what is now part of "critical habitat" for acroporid corals. On 9 May 2006, both the staghorn and elkhorn corals were placed on the threatened species list under the Endangered Species Act.

4. While conducting high-resolution seismic profiling in the area, the authors and shipboard colleagues were invited to dive on the *Atocha* while salvage operations were in progress. The senior author recovered several silver coins and a piece of the limestone upon which the ship grounded. It was oolite.

Chapter 5. Coral Health, or Lack Thereof

1. As of summer 2014, much of the centralized sewerage system had been installed on major Keys such as Marathon. As a result, building restrictions have been relaxed, and explosive demographic changes are in progress. Coral reefs continue to be in decline.

2. In the 1970s, new NOAA charts reversed names for Grecian Rocks and Key Largo Dry Rocks. Thus, in pre-1975 literature, what is now Grecian Rocks was called Key Largo Dry Rocks. That is true for the growth-rate study mentioned.

Chapter 6. Key Limes, Hands On: See It in Person

1. We do not recommend specific sunscreens because many contain chemicals that have been shown to be harmful to corals. A bulletin from the National Park Service states, "While no sunscreen has been proven to be completely 'reef-friendly,' those with titanium oxide or zinc oxide, which are natural mineral ingredients, have not been found harmful to corals."

2. Older field guides and publications call this reef Key Largo Dry Rocks. The reef to the north where the Christ statue resides was once called Grecian Rocks. Mapmakers reversed the names in the 1970s.

3. The Christ statue was installed before the area became a Federal Sanctuary. Because of separation of church and state, sanctuary personnel have no authority over condition of the bronze figure but do have protective authority for the marine life that grows on it. Nonetheless, some unknown individuals occasionally manage to clean the statue.

Epilogue: The Future for Reefs and Research in the Florida Keys

1. Accommodation is the space made available for potential sediment accumulation or vertical coral growth. Accommodation is a function of sea-level rise, subsidence, or a combination of both processes. Accommodation refers to all the space available, including old space (left over from an earlier time, such as empty Pleistocene backreef troughs) plus new space added, whereas new space added refers only to space contemporaneously being made available (such as at Looe Key Reef, for example, where sediment is being transported offshelf).

2. With each rise and fall of sea level, prehistoric peoples migrated north or south in response to fluctuations in landmass size and extent, which, in turn, altered availability of freshwater, animals, and botanicals to hunt for food. Freshwater and food sources also ebbed and flowed with the sea.

3. The youngest Pleistocene and oldest Holocene coral dates (77.8 ka of Multer et al., 2002, and 9.6 ka of Mallinson et al., 2003) bracket a period (~68 ka) when the outer Florida shelf was subaerially exposed; however, calcrete dates point to a prolonged period of subaerial exposure (~115 ka) for the Keys and mainland between Isotope Substage-5a time and the Holocene (figs. 1.1, 1.8A-B). Today, most dry land in the Florida Keys will disappear in a sea-level rise of 1–2 m (3–6.5 ft; Lidz and Shinn, 1991).

4. Loran refers to any of various long-range radio position-fixing systems by which hyperbolic lines of position are determined by measuring the difference in arrival times of synchronized pulse signals from two or more fixed transmitting radio stations of known geographic position.

REFERENCES CITED

Agassiz, A., 1883, Explorations of the surface fauna of the Gulf Stream, under the auspices of the United States Coast Survey, II. The Tortugas and Florida reefs: Memoir of the American Academy of Arts and Science Centennial, v. 2, p. 107–134, 12 pl.

Ball, M.M., Shinn, E.A., and Stockman, K.W., 1967, The geologic effects of Hurricane Donna in South Florida: Journal of Geology, v. 75, no. 5, p. 583–597.

Chappell, J., and Shackleton, N.J., 1986, Oxygen isotopes and sea level: Nature, v. 324, p. 137–140.

Davis, G.E., 1982, A century of natural change in coral distribution at the Dry Tortugas: A comparison of reef maps from 1881 and 1976: Bulletin of Marine Science, v. 32, no. 2, p. 608–623.

Drew, H.G., 1914, On the precipitation of calcium carbonate in the sea by marine bacteria, and on the action of denitrifying bacteria in tropic and temperate seas: Papers of the Tortugas Laboratory, Carnegie Institution of Washington Publication 182, p. 7–45.

Dunham, R.J., 1962, Classification of carbonate rocks according to depositional texture, *in* Ham, W.E., ed., Classification of Carbonate Rocks: American Association of Petroleum Geologists Memoir 1, p. 108–121.

Dunham, R.J., 1970, Keystone vugs in carbonate beach deposits (abstract): American Association of Petroleum Geologists Bulletin, v. 54, p. 845.

Eberli, G.P., McNeill, D.F., and Harris, P.M., 2014, Geology of the Everglades National Park and the Florida Keys: University of Miami American Association of Petroleum Geologists Student Chapter, Field Trip Guidebook, 120 p.

Enos P., 1977, Holocene sediment accumulations of the South Florida shelf margin, pt. I, *in* Enos, P., and Perkins, R.D., eds., Quaternary Sedimentation in South Florida: Geological Society of America Memoir 147, p. 1–130.

Fairbanks, R.G., 1989, A 17,000-year glacio-eustatic sea level record: Influence of glacial melting rates on the Younger Dryas event and deep ocean circulation: Nature, v. 342, p. 637–642.

Freode, C.R., Jr, and Shinn, E.A., 2012, Holocene beachrock in the Dry Tortugas, Florida, U.S.A.: Southeastern Geology, v. 49, no. 2, p. 79–90.

Garrison, V.H., Majewski, M.S., Foreman, W.T, Genualdi, S.A., Mohammed, A., and Massey Simonich., S.L., 2014, Persistent organic contaminants in Saharan dust air masses in West Africa, Cape Verde and the eastern Caribbean: Science of the Total Environment, v. 468–469, p. 530–543 (doi:10.1016/j.scitotenv.2013.08.076).

Ginsburg, R.N., 1953, Beachrock in south Florida: Journal of Sedimentary Petrology, v. 23, no. 2, p. 85–92.

Ginsburg, R.N., 1956, Environmental relationships of grain size and constituent particles in some South Florida carbonate sediments: American Association of Petroleum Geologists Bulletin, v. 40, no. 10, p. 2384–2427.

Ginsburg, R.N., 1964, South Florida Carbonate Sediments, Field Trip 1 Guide: Miami Beach, Geological Society of America Annual Meeting, 72 p.

Ginsburg, R.H., 1972, South Florida Carbonate Sediments: Comparative Sedimentology Laboratory, University of Miami, 71 p.

Ginsburg, R.N., and Lowenstam, H.A., 1958, The influence of marine bottom communities on the depositional environment of sediments: The Journal of Geology, v. 66, no. 3, p. 310–318.

Ginsburg, R.N., and Shinn, E.A., 1964, Distribution of the reef-building community in Florida and the Bahamas (abstract): American Association of Petroleum Geologists Bulletin, v. 48, p. 527.

Griffin, D.W., 2007, Atmospheric movement of microorganisms in clouds of desert dust and implications for human health: Clinical Microbiology Reviews, v. 20, no. 3, p. 459–477.

Harris, P.M., 1994, Satellite images and description of study areas, in Harris, P.M., and Kowalik, W.S., eds., Satellite Images of Carbonate Depositional Settings: Examples of Reservoir- and Exploration-Scale Geologic Facies Variation: American Association of Petroleum Geologists Methods in Exploration Series No. 11, p. 29–140 (posted in 2009 on AAPG Search and Discovery).

Harris, P.M., and Moore, C.H., 1985, Modern and ancient shelf carbonates—Facies, diagenesis, and petroleum potential: The Modern of South Florida and the Bahamas, The Ancient of the U.S. Gulf Coast: Colorado School of Mines Short Course Notes, 172 p.

Hine, A.C., 2013, Geologic History of Florida—Major Events That Formed the Sunshine State: Gainesville, University Press of Florida, 229 p.

Hine, A.C., Chambers, D.P., Clayton, T.D., Hafen, M.R., and Mitchum, G.T., 2016, Sea Level Rise in Florida: Science, Impacts, and Options: Gainesville, University Press of Florida, 178 p.

Hoffmeister, J.E., 1974, Land from the Sea, the Geologic Story of South Florida: Coral Gables, FL, University of Miami Press, 143 p.

Hoffmeister, J.E., and Ladd, H.S., 1944, The antecedent-platform theory: Journal of Geology, v. LII, no. 6.

Holmes, D.W., and Miller, R., 2004, Atmospherically transported elements and deposition in the Southeastern United States: Local or transoceanic? Applied Geochemistry, v. 19, p. 1189–1200.

Hudson, J.H., 1979, Absolute growth rates and environmental implications of Pleistocene *Montastrea annularis* in southeast Florida: Geological Society of America, Program with Abstracts, v. 11, no. 3, p. 85.

Hudson, J.H., 1981, Growth rates in *Montastraea annularis*: A record of environmental change in Key Largo Coral Reef Marine Sanctuary, Florida: Bulletin of Marine Science, v. 31, no. 2, p. 444–459.

Hudson, J.H., 1985, Growth rate and carbonate production in *Halimeda opuntia*, Marquesas Keys, Florida, in Toomey, D.F., and Nitecki, M.H., eds., Paleoalgology: Contemporary Research and Applications: Berlin, Heidelberg, Springer-Verlag, p. 257–263.

Hudson, J.H., Shinn, E.A., Halley, R.B., and Lidz, B.H., 1976, Sclerochronology—A tool for interpreting past environments: Geology, v. 4, p. 361–364.

Imbrie, J., Hays, J.D., Martinson, D.G., McIntyre, A., Mix, A.C., Morley, J.J., Pisias, N.G., Prell, W.L., and Shackleton, N.J., 1984, The orbital theory of Pleistocene climate: Support from a revised chronology of the marine $\delta^{18}O$ record, in Berger, A.L., Imbrie, J., Hays, J., Kukla, G., and Saltzman, B., eds., Milankovitch and Climate. Understanding the Response to Astronomical Forcing: Proceedings, NATO Advanced Research Workshop, Palisades, NY, NATO Science Series C, v. 126, p. 269–305.

Jaap, W.C., 1984, The ecology of the south Florida coral reefs: A community profile: U.S. Department of the Interior, Fish and Wildlife Service Report FWS/OBS-82/08 and Minerals Management Service Report MMS 84–0038, p. 1–138.

Kellogg, C.A., Griffin, D.W., and Shinn, E.A., 2002, Characterization of microbial communities associated with African desert dust and their implications for global human & ecosystem health: PAHO/WHO Conference & Workshop on Climate Variability and Change and their Health Effects in the Caribbean: Bridgetown, Barbados, West Indies.

Lessios, H.A., Robertson, D.R., and Cubit, J.D., 1984, Spread of *Diadema* mass mortalities through the Caribbean: Science, v. 226, p. 335–337.

Lidz, B.H., 2000a, Bedrock Beneath Reefs: the Importance of Geology in Understanding Biological Decline in a Modern Ecosystem: U.S. Geological Survey Open-File Report 00-046, 4 p., 1 pl.

Lidz, B.H., 2000b, Reefs, Corals, and Carbonate Sands: Guides to Reef-Ecosystem Health and Environment: U.S. Geological Survey Open-File Report 00-164, 6 p., 1 pl.

Lidz, B.H., 2005, Mapping Benthic Ecosystems and Environments in the Florida Keys: U.S. Geological Survey Open-File Report 2005-1002, 6 p., 1 pl.

Lidz, B.H., 2006, Pleistocene corals of the Florida Keys: Architects of imposing reefs—Why?: Journal of Coastal Research, v. 22, no. 4, p. 750–759.

Lidz, B.H., Brock, J.C., and Nagle, D.B., 2008b, Utility of shallow-water ATRIS images in defining biogeologic processes and self-similarity in skeletal Scleractinia, Florida reefs: Journal of Coastal Research, v. 24, no. 5, p. 1320–1338.

Lidz, B.H., and Hallock, P., 2000, Sedimentary petrology of a declining reef ecosystem, Florida reef tract (U.S.A.): Journal of Coastal Research, v. 16, no. 3, p. 675–697.

Lidz, B.H., Hine, A.C., Shinn, E.A., and Kindinger, J.L., 1991, Multiple outer-reef tracts along the South Florida bank margin: Outlier reefs, a new windward-margin model: Geology, v. 19, p. 115–118.

Lidz, B.H., Reich, C.D., Peterson, R.L., and Shinn, E.A., 2006, New maps, new information: Coral reefs of the Florida Keys: Journal of Coastal Research, v. 22, no. 2, p. 61–83.

Lidz, B.H., Reich, C.D., and Shinn, E.A., 2003, Regional Quaternary submarine geomorphology in the Florida Keys: Geological Society of America Bulletin, v. 115, no. 7, p. 845–866, plus oversize color plate of Pleistocene topographic and Holocene isopach maps.

Lidz, B.H., Reich, C.D., and Shinn, E.A., 2007, Systematic mapping of bedrock and habitats along the Florida reef tract: Central Key Largo to Halfmoon Shoal (Gulf of Mexico): U.S. Geological Survey Professional Paper 1751, http://pubs.usgs.gov/pp/2007/1751.

Lidz, B.H., and Shinn, E.A., 1991, Paleoshorelines, reefs, and a rising sea: South Florida, U.S.A: Journal of Coastal Research, v. 7, no. 1, p. 203–229.

Lidz, B.H., Shinn, E.A., Hine, A.C., and Locker, S.D., 1997, Contrasts within an outlier-reef system: Evidence for differential Quaternary evolution, South Florida windward margin, U.S.A.: Journal of Coastal Research, v. 13, no. 3, p. 711–731.

Lidz, B.H., Shinn, E.A., Hudson, J.H., Multer, H.G., Halley, R.B., and Robbin, D.M., 2008a, Controls on late Quaternary coral reefs of the Florida Keys, p. 9–74 *in* Riegl, B.M., and Dodge, R.E., eds., Coral Reefs of the World Vol. 1, Coral Reefs of the USA: Berlin, Springer Science + Business Media B.V., 803 p.

Lidz, B.H., and Zawada, D.G., 2013, Possible return of *Acropora cervicornis* at Pulaski Shoal, Dry Tortugas National Park, Florida: Journal of Coastal Research, v. 29, no. 2, p. 256–271 (doi: 10.2112/JCOASTRES-D-12–00078.1).

Lighty, R.G., Macintyre, I.G., and Stuckenrath, R., 1982, *Acropora palmata* reef framework: A reliable indicator of sea level in the western Atlantic for the past 10,000 years: Coral Reefs, v. 1, p. 125–130.

Locker, S.D., Hine, A.C., Tedesco, L.P., and Shinn, E.A., 1996, Magnitude and timing of episodic sea-level rise during the last deglaciation: Geology, v. 24, no. 9, p. 827–830.

Ludwig, K.R., Muhs, D.R., Simmons, K.R., Halley, R.B., and Shinn, E.A., 1996, Sea-level records at 80 ka from tectonically stable platforms: Florida and Bermuda: Geology, v. 24, no. 3, p. 211–214.

Macintyre, I.G., 1975, A diver-operated hydraulic drill for coring submerged substrates: Atoll Research Bulletin, v. 185, p. 21–26.

Mallinson, D., Hine, A., Hallock, P., Locker, S., Shinn, E., Naar, D., Donahue, B., and Weaver, D., 2003, Development of small carbonate banks on the South Florida platform margin: Response to sea level and climate change: Marine Geology, v. 199, p. 45–63.

Mayer, A.G., 1903, The Tortugas, Florida, as a station for research in biology: Science, v. 17, no. 422, p. 190–192.

Muhs, D.R., Bush, C.A., and Stewart, K.C., 1990, Geochemical evidence of Saharan dust parent material for soils developed on Quaternary limestones of Caribbean and Western Atlantic islands: Quaternary Research, v. 33, p. 157–177.

Multer, H.G., 1974, Field Guide to Some Carbonate Rock Environments, Florida Keys-Western Bahamas: Guide Book for Geological Society of America Annual Meeting, Miami, FL, and Published by the Author, 430 p.

Multer, H.G., 1977, Field Guide to Some Carbonate Rock Environments: Florida Keys and Western Bahamas: Dubuque, Iowa, Kendall/Hunt, 435 p.

Multer, H.G., Gischler, E., Lundberg, J., Simmons, K.R., and Shinn, E.A., 2002, Key Largo Limestone revisited: Pleistocene shelf-edge facies, Florida Keys, USA: Facies, v. 46, p. 229–272.

Multer, H.G., and Hoffmeister, J.E., 1968, Subaerial laminated crusts of the Florida Keys: Geological Society of America Bulletin, v. 79, p. 183–192.

Paul, J.H., Rose, J.B., Brown, J., Shinn, E.A., Miller, S., and Farrah, S.R., 1995a, Viral tracer studies indicate contamination of marine waters by sewage disposal practices in Key Largo, Florida: Applied and Environmental Microbiology, v. 61, no. 6, p. 2230–2234.

Paul, J.H., Rose, J.B., Jiang S., Kellogg, C., and Shinn, E.A., 1995b, Occurrence of fecal indicator bacteria in surface waters and the subsurface aquifer in Key Largo, Florida: Applied and Environmental Microbiology, v. 61, no. 6, p. 2235–2241.

Perkins, R.D., 1977, Depositional framework of Pleistocene rocks in South Florida, pt. II, *in* Enos, P., and Perkins, R.D., eds., Quaternary Sedimentation in South Florida: Geological Society of America, Memoir 147, p. 131–198.

Perkins, R.D., and Enos, P., 1968, Hurricane Betsy in the Florida-Bahamas area: Geologic effects and comparison with Hurricane Donna: Journal of Geology, v. 76, p. 710–717.

Porter, J.W., Lewis, S.K., and Porter, K.G., 1999, The effect of multiple stressors on the Florida Keys coral reef ecosystem: A landscape hypothesis and a physiological test: Limnology and Oceanography, v. 44, p. 941–949.

Precht, W.F., and Aronson, R.B., 2004, Climate flickers and range shifts of reef corals: Frontiers in Ecological Environments, v. 2, p. 307–314.

Precht, W.F., and Miller, S.L., 2007, Ecological shifts along the Florida reef tract: The past as a key to the future, *in* Aronson, R.B., ed., Geological Approaches to Coral Reef Ecology: New York, Springer-Verlag, p. 237–312.

Prospero, J.M., 1999, Long-term measurements of the transport of African mineral dust to the southeastern United States: Implications for regional air quality: Journal of Geophysical Research, v. 104, p. 15,917–15,927.

Prospero, J.M., and Nees, R.T., 1986, Impact of the North African drought and El Niño on mineral dust in the Barbados trade winds: Nature, v. 320, p. 735–738.

Purdy, E.G., 1974, Karst-determined facies patterns in British Honduras: Holocene carbonate sedimentation model: American Association of Geologists Bulletin, v. 58, p. 825–855.

Reich, C.D., 1996, Diver-operated manometer: A simple device for measuring hydraulic head in underwater wells: Journal of Sedimentary Research, v. 66, no. 5, p. 1032–1034.

Reich, C.D., Shinn, E.A., Hickey, T.D, and Tihansky, A.B., 2002, Tidal and meteorological influences on shallow marine groundwater flow in the upper Florida Keys, in Porter, J.W., and Porter, K.G., eds., The Everglades, Florida Bay, and Coral Reefs of the Florida Keys: Boca Raton, FL, CRC Press, p. 659–676.

Risebrough, R.W., Huggett, R.J., Griffin, J.J., and Goldberg, E.D., 1968, Pesticides: Transatlantic movement in the northeast trades: Science, v. 159, p. 1233–1235.

Robbin, D.M., 1981, Subaerial $CaCO_3$ crust: A tool for timing reef initiation and defining sea level changes: International Coral Reef Symposium, 4th, Manila, Philippines, Proceedings, v. 1, p. 575–579.

Robbin, D.M., 1984, A new Holocene sea level curve for the upper Florida Keys and Florida reef tract, in Gleason, P.J., ed., Environments of South Florida: Present and Past, 2nd ed.: Miami, Miami Geological Society, Memoir 2, p. 437–458.

Robbin, D.M., and Stipp, J.J., 1979, Depositional rate of laminated soilstone crusts, Florida Keys: Journal of Sedimentary Petrology, v. 49, p. 0175–0180.

Roberts, H.H., Rouse, L.J., Jr., Walker, N.D., and Hudson, J.H., 1982, Cold-water stress in Florida Bay and northern Bahamas—A product of winter cold-air outbreaks: Journal of Sedimentary Petrology, v. 52, p. 145–155.

Romans, B., 1775 (introduction by R.W. Patrick, 1962), A Concise Natural History of East and West Florida, a Facsimile Reproduction of the 1775 Edition: Gainesville, University of Florida Press, 340 p.

Scholl, D.W., Craighead, F.C., Sr., and Stuiver, M., 1969, Florida submergence curve revised: Its relation to coastal sedimentation rates: Science, New Series, v. 163, no. 3867, p. 562–564.

Scholle, P.A., 1978, A color illustrated guide to carbonate rock constituents, textures, cements and porosities: American Association of Petroleum Geologists Memoir 27, 241 p.

Shackleton, N.J., 1987, Oxygen isotopes, ice volume and sea level: Quaternary Science Review, v. 6, p. 183–190.

Shinn, E.A., 1963, Spur and groove formation on the Florida reef tract: Journal of Sedimentary Petrology, v. 33, p. 291–303.

Shinn, E.A., 1966, Coral growth rate, an environmental indicator: Journal of Paleontology, v. 40, p. 233–240.

Shinn, E.A., 1968, Burrowing in Recent lime sediments of Florida and the Bahamas: Journal of Paleontology, v. 42, no. 4, p. 879–894.

Shinn, E.A., 1976, Coral reef recovery in Florida and the Persian Gulf: Environmental Geology, v. 1, p. 241–254.

Shinn, E.A., 1980, Geologic history of Grecian Rocks, Key Largo Coral Reef Marine Sanctuary: Bulletin of Marine Science, v. 30, p. 646–656.

Shinn, E.A., 1984, Geologic history, sediment, and geomorphic variations within the Florida reef tract: Symposium on Advances in Reef Science, a joint meeting of the Atlantic Reef Committee, Rosenstiel School of Marine and Atmospheric Science, University of Miami, and the International Society for Reef Studies, Miami, Abstracts of Papers, p. 113–114.

Shinn, E.A., 2011, Spurs and grooves, in Hopley, D., ed., Encyclopedia of Modern Coral Reefs, Structure, Form and Processes: Encyclopedia of Earth Sciences Series, Springer, p. 1032–1034.

Shinn, E.A., 2013, Bootstrap Geologist: My Life in Science: Gainesville, University Press of Florida, 297 p.

Shinn, E.A., Halley, R.B., and Hine, A.C., 2000b, SEPM Field Guide to the Florida Reef Tract, Key Largo Area: Tulsa, OK, SEPM (Society for Sedimentary Geology), 47 p.

Shinn, E.A., Hudson, J.H., Halley, R.B., and Lidz, B.H., 1977, Topographic control and accumulation rate of some Holocene coral reefs, South Florida and Dry Tortugas, *in* International Coral Reef Symposium, 3rd, Miami, Proceedings, v. 2, Geology, p. 1–7.

Shinn, E.A., and Jaap, W.C., 2011, Field Guide to the Major Organisms and Processes Building Reefs and Islands of the Dry Tortugas: The Carnegie Dry Tortugas Laboratory Centennial Celebration (1905–2005): Southeastern Geological Society Field Guide #54, 43 p. (Guide was written in 2005 for a field trip conducted that year but not printed for general use until 2011).

Shinn, E.A., and Lidz, B.H., 1988, Blackened limestone pebbles: Fire at subaerial unconformities, *in* James, N.P., and Choquette, P.W., eds., Paleokarst: New York, Springer-Verlag, p. 117–131.

Shinn, E.A., Lidz, B.H., and Holmes, C.W., 1990, High-energy carbonate sand accumulation, The Quicksands, southwest Florida Keys: Journal of Sedimentary Petrology, v. 60, no. 6, p. 952–967.

Shinn, E.A., Lidz, B.H., Kindinger, J.L., Hudson, J.H., and Halley, R.B., 1989, Reefs of Florida and the Dry Tortugas: A guide to the modern carbonate environments of the Florida Keys and the Dry Tortugas: International Geological Congress Field Trip Guidebook T176: Washington, DC, American Geophysical Union, 55 p.

Shinn, E.A., Reese, R.S., and Reich, C.D., 1994, Fate and Pathways of Injection-Well Effluent in the Florida Keys: U.S. Geological Survey, Open-File Report 94–276, 116 p.

Shinn, E.A., Reich, C.D., Hickey, D.T., and Lidz, B.H., 2003, Staghorn tempestites in the Florida Keys: Coral Reefs, v. 22, p. 91–97.

Shinn, E.A., Reich, C.D., Locker, S.D., and Hine, A.C., 1996, A giant sediment trap in the Florida Keys: Journal of Coastal Research, v. 12, no. 4, p. 953–959.

Shinn, E.A., Robbin, D.M., and Claypool, G.E., 1984, Compaction of modern carbonate sediments: Implications for generation and expulsion of hydrocarbons, *in* Palacas, J.E., ed., Petroleum Geochemistry and Source Rock Potential of Carbonate Rocks: American Association of Petroleum Geologists Studies in Geology #18, p. 197–203.

Shinn, E.A., Smith, G.W., Prospero, J.M., Betzer, P., Hayes, M.L., Garrison, V., and Barber, R.T., 2000a, African dust and the demise of Caribbean coral reefs: Geophysical Research Letters, v. 27, no. 19, p. 3029–3032.

Swap, R., Garstang, M., Greco, S., Talbot, R., and Kallberg, P., 1992, Saharan dust in the Amazon Basin: Tellus, v. 44B, no. 2, p. 133–149.

Swinchatt, J.P., 1965, Significance of constituent composition, texture, and skeletal breakdown in some Recent carbonate sediments: Journal of Sedimentary Petrology, v. 35, no. 1, p. 71–90.

Turmel, R.J., and Swanson, R.G., 1976, The development of Rodriguez Bank, a Holocene mudbank in the Florida reef tract: Journal of Sedimentary Petrology, v. 46, no. 3, p. 497–518.

Vaughan, T.W., 1914, Building of the Marquesas and Tortugas atolls and a sketch of the geologic history of the Florida reef tract: Carnegie Institution of Washington Publication 182, Papers of the Department of Marine Biology, v. 5, p. 55–67.

Vaughan, T.W., 1915a, Growth rate of the Floridian and Bahamian shoal-water corals: Carnegie Institution of Washington Yearbook No. 13, p. 221–231.

Vaughan, T.W., 1915b, The geological significance of the growth rate of the Floridan and Bahamian shoal water corals: Washington Academy of Science Journal, v. 5, p. 591–600.

Walsh, J.J., and Steidinger, K.A., 2001, Saharan dust and Florida red tides: The cyanophyte connection: Journal of Geophysical Research, v. 106, p. 11,597–11,612.

Wanless, H.R., 1981, Fining upward sedimentary sequences generated in seagrass beds: Journal of Sedimentary Petrology, v. 51, no. 2, p. 445–454.

Wanless, H.R., 1989, The inundation of our coastlines: Past, present and future with a focus on South Florida: Sea Frontiers, v. 35, no. 5, p. 264–271.

Wanless, H.R., and Tagett, M.G., 1989, Origin, growth, and evolution of carbonate mudbanks in Florida Bay: Bulletin of Marine Science, v. 44, p. 454–489.

Wanless, H.R., and Tedesco, L.P., 1993, Depositional and early diagenetic controls on texture and fabric of carbonate mudbanks, south Florida, *in* Rezak, R., and Lavoie, D.L., eds., Carbonate Microfabrics, Frontiers in Sedimentary Geology: New York, Springer-Verlag, p. 41–63.

Young, R.W., Carder, K.L., Betzer, P.R., Costello, D.K., Duce, R.A., DiTullio, G.R., Tindale, N.W., Laws, E.A., Uematsu, M., Merrill, J.T., and Feely, R.A., 1991, Atmospheric iron inputs and primary productivity: Phytoplankton responses in the North Pacific: Global Biogeochemistry Cycles, v. 5, no. 2, p. 119–134.

INDEX

The authors are known worldwide for their published scientific findings. Eugene Shinn recently retired from his position as a research geologist with the U.S. Geological Survey (USGS) Coastal and Marine Science Center in St. Petersburg, Florida. Currently he is a courtesy professor at the University of South Florida, College of Marine Science. After obtaining his Bachelor of Science degree from the University of Miami, Shinn spent a year at the University's Marine Laboratory. In 1958, he began working for Shell Oil Company as a research geologist. During his 15-year career with Shell, he worked in Florida and the Bahamas, the Persian Gulf, and Holland. He completed that career at Shell Headquarters in Houston, TX in the environmental affairs department. Gene joined the USGS in 1974 and established the USGS Fisher Island Field Station off Miami, which he directed for 15 years. For more information, see his memoir/autobiography, *Bootstrap Geologist: My Life in Science* (University Press of Florida, 2014). Shinn is a GSA Fellow, former VP of SEPM, Honorary Member of SEPM, and SEPM Twenhofel medalist, and he has an Honorary PhD from the University of South Florida.

Barbara Lidz is a scientist emerita research geologist with the USGS Coastal and Marine Science Center in St. Petersburg. A Smith College alumna, she obtained her Bachelor of Science degree from the University of Miami and spent 12 years at the University's Marine Laboratory before joining the USGS in 1974. Her project involvement and studies have included seismic data collection throughout the Florida Keys, oil well searches off the west coast of Florida and off Key West, and seismic-profiling cruises within the Gulf of Mexico, Caribbean, Bahamas, and Florida Keys. She is known for her many publications, ranging from the micropaleontology and biostratigraphy of deep-sea sediments on the Great Bahama Bank and in the Caribbean to the geologic history of Florida's coral reefs. She is coauthor or lead author on dozens of papers with the book's lead author. Lidz is a Fellow of the Cushman Foundation for Foraminiferal Research and served as Editor and Chairman of SEPM Publications and as Executive Officer on the SEPM Council for eight years. She is an Honorary Member of the Miami Geological Society, having served in all four official capacities including President, plus as Publications Editor and member of the Board of Directors. Lidz was Editor of the USGS Bureau-wide Coastal and Marine Science newsletter *Sound Waves* for 15 years.